MATHEMATICAL WORLD • VOLUME 24

The Shoelace Book

A Mathematical Guide
to the Best (and Worst) Ways
to Lace Your Shoes

Burkard Polster

AMERICAN MATHEMATICAL SOCIETY

For

Dudu and Joujou

2000 *Mathematics Subject Classification.* Primary 00A05, 90C27, 05A15.

The photograph on the cover was taken by the author.

For additional information and updates on this book, visit
www.ams.org/bookpages/mawrld-24

Library of Congress Cataloging-in-Publication Data
Polster, Burkard.
   The shoelace book : a mathematical guide to the best (and worst) ways to lace your shoes /
Burkard Polster.
      p. cm. — (Mathematical world, ISSN 1055-9426 ; v. 24)
   Includes bibliographical references and index.
   ISBN 0-8218-3933-0 (softcover : alk. paper)
   1. Mathematics—Miscellanea.   2. Combinatorial optimization.   3. Shoelaces—Mathematics.
I. Title.   II. Series.

QA99.P65   2006
510—dc22
                                                                                2006040733

Contents

Preface

Have you ever had problems with your shoelaces? With broken shoelaces? With shoelaces that constantly come undone? Tripped over your shoelaces? No, that is not what I mean. What I have in mind are mathematical problems such as the following:

- What is the shortest way to lace your shoes?
- What is the strongest way to lace your shoes?
- How many ways are there to lace your shoes?

On December 5, 2002, a short article of mine [21] that addresses these questions appeared in the journal *Nature*. I don't think anybody was more surprised than I by the incredible amount of publicity attracted by this note on as innocent a topic as shoelaces. In the weeks following its publication, the article was reported on by virtually every major newspaper worldwide, and I received close to one thousand e-mails in which people from all walks of life asked me about the mathematics of shoelaces. This is even more remarkable since the article was not even one page long and merely contained a summary of some of my answers to the above questions without any proofs. This set of notes has been compiled in an attempt to provide the comprehensive account of shoelace mathematics that many people have asked me for.

To start with, pondering questions about the mathematics of shoelaces was not much more than idle doodling on my part. It soon became clear to me that other mathematicians had already thought about the shortest shoelace problem and had come up with some very complete and neat theorems, arrived at via conceptually appealing proofs; see [13], [14], [16], and [20]. This initial trend towards beautiful results continued throughout my subsequent investigations, and, in the end, a very complete picture emerged, consisting mainly of simple, beautiful, and often surprising characterizations of the most common shoelace patterns, arrived at via elementary, yet pretty and nonobvious, mathematics. I think such a picture is worth painting in detail, as many mathematically minded people will be interested in it for as long as they use shoelaces to tie their shoes.

Summary of Contents

In Chapter 1, we collect the most basic definitions and results about n-lacings of a mathematical shoe, the mathematical counterparts of lacings of a shoe with n pairs of eyelets. Here, a mathematical shoe consists of $2n$ eyelets which are the points of intersection of two vertical lines in the plane that are one unit apart and n equally spaced horizontal lines that are h units apart. An n-lacing of a mathematical shoe is a closed path in the plane consisting of $2n$ line segments whose endpoints are the $2n$ eyelets of the shoe. Furthermore, we require that given any eyelet E, at least one of the two segments ending in it is not contained in the same column of eyelets as E. This condition ensures that every eyelet genuinely contributes towards pulling the two sides of the shoe together or, less formally, that a lacing doesn't have "gaps".

We introduce four important special classes of n-lacings. The *dense* n-lacings are the n-lacings in which the shoelace zigzags back and forth between the two columns of eyelets. The *straight* n-lacings are those n-lacings that contain all possible horizontal segments. The *superstraight* n-lacings are the straight n-lacings all of whose nonhorizontal segments are verticals. Finally, if, when you trace an n-lacing, you move exactly once from the top to the bottom of the shoe and once from the bottom to the top and if you neither "backtrack" on the way down nor on the way up, then the n-lacing is called *simple*.

We describe some families of n-lacings, representatives of which are actually used for lacing real shoes and which pop up in the different characterizations that this set of notes is all about. See the diagram on the next page for a quick visual description of these and other important families of n-lacings and a summary of the most important such characterizations. For example, the two most commonly used n-lacings are the so-called crisscross and zigzag n-lacings, which are featured on the left side of the diagram. As you can see, they both have very neat extremal properties.

In Chapter 2, we consider one-column n-lacings. Imagine pulling really hard on the two ends of the shoelace in one of your shoes that has been laced using a straight lacing. Then, if the lacing does not get in the way and if your foot is narrow enough, you will end up with the two columns of eyelets superimposed, one on top of the other. This means that we do not have to distinguish any longer between the two columns of eyelets and what we are dealing with is a one-column n-lacing, that is, a lacing of just one column of n eyelets in which every eyelet gets visited exactly once. We identify the shortest and longest one-column n-lacings and find the numbers of such lacings. In a final section, we describe a simple method that allows us to construct all straight n-lacings that contract to a given one-column n-lacing. This method plays an important role in deriving the shortest and longest straight n-lacings in subsequent chapters.

In Chapter 3, we derive one formula each for the numbers of those n-lacings that belong to one of ten different classes of n-lacings considered by us: general, dense, simple, straight, dense-and-simple, dense-and-straight, etc. The highlights of this chapter are the formula for the number of n-lacings and the formula for the number of simple n-lacings. Especially, the latter is a very striking example of a simple mathematical object giving rise to a beautiful, yet surprisingly complicated, formula. Also included in this chapter are complete lists of all 2-lacings, all 3-lacings, and all simple 4-lacings.

In Chapter 4, we extend results by Halton [13] and Isaksen [16] by deriving the shortest n-lacings in the different classes of n-lacings in which we are interested. Highlights of this chapter are our proofs that the bowtie n-lacings are the shortest n-lacings overall, that the crisscross n-lacings are the shortest dense n-lacings, and

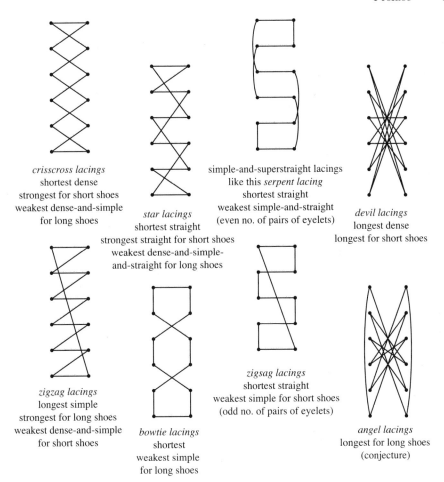

crisscross lacings
shortest dense
strongest for short shoes
weakest dense-and-simple
for long shoes

star lacings
shortest straight
strongest straight for short shoes
weakest dense-and-simple-
and-straight for long shoes

simple-and-superstraight lacings
like this serpent lacing
shortest straight
weakest simple-and-straight
(even no. of pairs of eyelets)

devil lacings
longest dense
longest for short shoes

zigzag lacings
longest simple
strongest for long shoes
weakest dense-and-simple
for short shoes

bowtie lacings
shortest
weakest simple
for long shoes

zigsag lacings
shortest straight
weakest simple for short shoes
(odd no. of pairs of eyelets)

angel lacings
longest for long shoes
(conjecture)

that the star n-lacings are the shortest among all those n-lacings that are both dense and straight.

In Chapter 5, we consider variations of the shortest shoelace problem. For example, we derive the solution of the shortest shoelace problem for lacings that are not closed, for lacings consisting of more than one shoelace, and for lacings of shoes in which the eyelets are not perfectly aligned. In particular, generalizing a result by Misiurewicz [20], we show that the crisscross n-lacing of a general n-shoe is the shortest dense n-lacing of this shoe. This last result demonstrates that the solution of the shortest shoelace problem in the class of dense n-lacings is very robust, as it stays unchanged even when the underlying array of eyelets is perturbed quite radically. On the other hand, we also show that the same cannot be said about the bowtie n-lacings as the shortest n-lacings overall.

In Chapter 6, we derive the longest n-lacings in most of the different classes of n-lacings in which we are interested. One of the most surprising and interesting results of this chapter is the characterization of the zigzag n-lacings as the longest simple n-lacings.

In Chapter 7, we consider n-lacings as pulley systems and succeed in identifying the strongest pulleys in all the different classes of n-lacings in which we are interested. Most importantly, we prove that the crisscross and zigzag n-lacings are the strongest pulley systems among all n-lacings and that the star and zigzag n-lacings are the strongest straight n-lacings.

In Chapter 8, we derive the weakest n-lacings in some of the classes of n-lacings in which we are interested. This results in a number of new characterizations of the bowtie, crisscross, zigzag, zigsag, and star n-lacings.

In Appendix A, we first give a brief introduction to the so-called traveling salesman problems. These problems are close relatives of our shortest shoelace problems. We also describe the so-called shoelace formula for calculating the area of polygons.

In Appendix B, we collect all kinds of curious and interesting facts about real shoelaces and lacings.

Acknowledgements

I am grateful to Allen Offer and Hendrik Van Maldeghem, whose insightful questions prompted me to investigate simple and straight lacings. Eventually, this led to some further nice characterizations of some popular lacing methods. I would like to thank Ina Lindemann for her marvelous support of this project, Michael Eastwood for drawing my attention to the analysis of the relative strengths of different suturing methods in [22], and Marty Ross for his feedback on an earlier version of the manuscript. Finally, my most heartfelt thanks go to my wife, Anu, for accompanying me on countless shoelace-related expeditions.

Melbourne, Australia   Burkard Polster

1

Setting the Stage

We start by describing the simple models of shoes and lacings of shoes that we will be working with in the following.

A *mathematical shoe* consists of $2n$ *eyelets* which are the points of intersection of two vertical lines and n equally spaced horizontal lines in the plane. In everything that follows $n \geq 2$. We fix the distance between the two vertical lines to be 1 and call the distance h between two adjacent horizontal lines the *stretch* of the shoe; see Figure 1.1. We call the set of all eyelets contained in one of the vertical lines a *column of eyelets* and the set of two eyelets contained in a horizontal line a *row of eyelets*. We label the left column with an A, the right column with a B, and the rows of eyelets from 1 to n, proceeding from top to bottom. This induces a labelling of the eyelets.

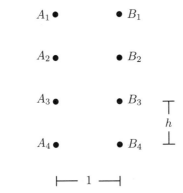

Figure 1.1. A mathematical shoe with stretch h.

An *n-lacing* of our mathematical shoe is a closed path in the plane consisting of $2n$ line *segments* whose endpoints are the $2n$ eyelets of the shoe. Furthermore, we require that, given any eyelet E, at least one of the two segments ending in it is not contained in the same column as E. This condition ensures that every eyelet genuinely contributes towards pulling the two sides of the shoe together or, less formally, that lacings don't have "gaps". Virtually all lacings that are actually used satisfy this condition. Note that our condition is equivalent to the following: As we trace along a lacing, no three consecutive eyelets visited are contained in the same column. The diagram in Figure 1.2 does not correspond to a lacing, as both segments that end in one of the middle eyelets are contained in the same column as this eyelet.

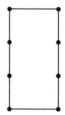

Figure 1.2. Not a lacing.

We call a segment of a lacing a *vertical* if its endpoints are both contained in column A or both in column B. We call a segment a *diagonal* if it is not a vertical. A diagonal is a *horizontal* if its endpoints have the same index, that is, are contained in the same row. The *vertical length* of a segment is the nonnegative difference between the indices of its endpoints. For example, the vertical length of a horizontal is 0. A segment is an *m-segment* if its vertical length is m. For example, a 0-diagonal is the same as a horizontal. The *length* of a lacing is the sum of the lengths of its segments.

It is important to realize that a picture of a lacing can correspond to more than one lacing. For example, consider Figure 1.3. The left- and right-hand diagrams represent two different "loose" lacings that both contract to the middle diagram when the shoelace is pulled taut. This means that, just by looking at the diagram in the middle, we cannot be sure which lacing it represents. To avoid this kind of ambiguity, we draw lacings loosely, whenever this is necessary.

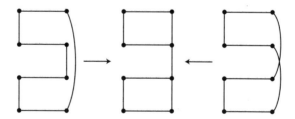

Figure 1.3. Two "loose" lacings that contract to the same diagram.

1.1 Popular Lacings

Figure 1.4 shows one representative each from six popular families of lacings that are actually used for various purposes. These are the *crisscross, zigzag, star, bowtie, serpent*, and *zigsag lacings*.

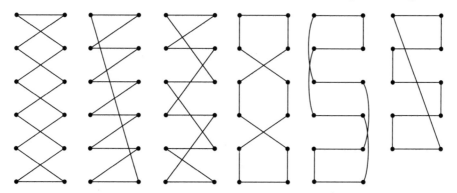

Figure 1.4. Representatives of six popular families of lacings: the crisscross, zigzag, star, bowtie, serpent, and zigsag lacings.

These lacings will play important roles in the following chapters, and it may therefore be a good idea to memorize them right now.[1]

Of course, the crisscross and the zigzag lacings are by far the most popular lacings today. Therefore, we will also refer to them as the *classical lacings*.

The crisscross, zigzag, and star n-lacings exist for all n. For $n = 2$ these three types of n-lacings coincide. For $n > 2$ there is a unique crisscross n-lacing, two zigzag n-lacings, and two star n-lacings. The two zigzag n-lacings are vertical mirror images of each other. Similarly, the two star n-lacings are vertical mirror images of each other. In all three cases, it should be clear from the diagrams how to construct their n-lacing representatives.

Figure 1.5 shows the basic building blocks of the bowtie lacings. From left to right, we have an *end*, a *cross*, and a *gap*. Here, an end is a horizontal that connects either the top two or the bottom two eyelets. This means that a lacing contains at most two ends. A cross consists of two 1-diagonals that are situated with respect to each other as shown in the middle diagram. Finally, a gap consists of two 1-verticals that are situated with respect to each as shown in the right-hand diagram.

Figure 1.5. The basic building blocks of the bowtie lacings.

Clearly, the crisscross n-lacing is made up of the two ends and $n-1$ crosses. An n-lacing is a bowtie n-lacing if it is made up of two ends and

- $n/2 - 1$ crosses and $n/2$ gaps if n is even, and
- $(n - 1)/2$ crosses and $(n - 1)/2$ gaps if n is odd.

---

[1] To facilitate memorization, we have chosen names that capture the construction principles behind the different lacings: (1) crisscross–note all the crosses in the middle of these lacings; (2) zigzag–discard the long segment and you end up with a long zigzag; (3) star–note the partial Stars of David that are hiding in these lacings; (4) bowtie–turn this bowtie 6-lacing 90 degrees and you see a bowtie; (5) serpent–this lacing winds left and right like a serpent; (6) zigsag–reminiscent of zigzag lacings.

It is an easy exercise to check that there is exactly one bowtie n-lacing if n is even and exactly $(n+1)/2$ different bowtie n-lacings in the case that n is odd. Figure 1.6 shows the different bowtie n-lacings for $n = 2, 3, 4, 5$, and 6.

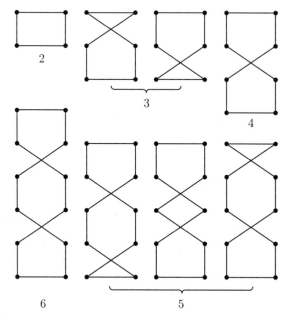

Figure 1.6. The bowtie n-lacings for $n = 2, 3, 4, 5$, and 6.

Note that for fixed n all bowtie n-lacings have the same length. Furthermore, by Lemma 1.1 below, a bowtie n-lacing contains the maximum number of verticals.

The serpent n-lacings clearly exist only for even n and the zigsag n-lacings exist only for odd n. For $n > 2$ both serpent and zigsag n-lacings come in pairs, one member of the pair being the vertical mirror image of the second. For $n = 2$ the bowtie and serpent n-lacings coincide. It should be clear from the diagram how to construct the n-lacing representatives of the serpent and zigsag lacings.

1.2 Dense, Straight, Superstraight, and Simple Lacings

We call a lacing *dense* if it does not contain any verticals. This means that a dense lacing zigzags back and forth between the two columns of eyelets. The crisscross, zigzag, and star n-lacings are dense, whereas the bowtie, serpent, and zigsag n-lacings are not.

We call an n-lacing *straight* if it contains all horizontals. Examples of straight n-lacings are the zigzag, star, serpent, and zigsag lacings. We call an n-lacing *superstraight* if it is straight and all nonhorizontal segments are verticals. The serpent n-lacings are examples of superstraight n-lacings.

Clearly, we can travel from eyelet A_1 to eyelet A_n along an n-lacing in exactly two different ways; see Figure 1.7. Let's note the indices of the eyelets we come across during one of these journeys in the order that we visit them. We call an n-lacing *simple* if the sequences of numbers corresponding to the associated journeys are both nondecreasing. All the n-lacings we have discussed so far are simple n-lacings. The lacing in Figure 1.7 is an example of a 3-lacing that is not simple.

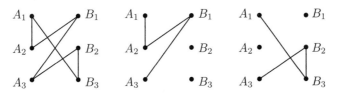

Figure 1.7. A 3-lacing that is not simple and the two ways to travel from A_1 to A_3.

We end this chapter with three lemmas that summarize some basic properties of n-lacings.

Lemma 1.1 (Verticals). *Let l be an n-lacing. Then the following hold:*

1. *The number of verticals of l in column A equals the number of verticals in column B.*
2. *The maximal possible number of verticals in l is n for n even, and $n-1$ for n odd.*

Proof. Let a, b, and d denote the number of verticals in column A, the number of verticals in column B, and the number of diagonals, respectively. Remember that any of the eyelets can be endpoint of *at most* one vertical and that any of the eyelets is endpoint of *at least* one diagonal. Therefore, there are exactly $2a$ eyelets in column A in which exactly one diagonal ends and $n - 2a$ eyelets in which exactly two diagonals end. Consequently, we can count the number of diagonals as follows:

$$2a + 2(n - 2a) = d = 2b + 2(n - 2b).$$

Hence, $a = b$. This proves the first part of this lemma. To prove the rest is a straightforward exercise.   □

Lemma 1.2 (Existence of Superstraight Lacings). *Superstraight n-lacings exist only for even n.*

Proof. The serpent n-lacings can be realized for any even n, which means that superstraight n-lacings exist for even n. A superstraight n-lacing contains $2n$ segments, n of which are horizontals and the remaining n are verticals. As a consequence of the previous lemma, this number of verticals is even.   □

Lemma 1.3 (Horizontals in Simple Lacings). *Every simple n-lacing contains the top horizontal and the bottom horizontal.*

This lemma is an immediate consequence of the definition of simple n-lacings.

1.3 Notes

Names. The different popular lacings are known under various names. For example, Halton [14] calls the crisscross lacings *American lacings*, the zigzag lacings *shoe shop lacings*, and the star lacings *European lacings*. The zigzag lacings are also sometimes referred to as *ladder lacings*, *spiral lacings*, or *straight lacings*. Isaksen [16] refers to the bowtie lacings as *iceskater lacings* and to the serpent lacings as *bowling lacings*.

Other lacings. Various authors have considered lacings of shoes different from those considered by us. In particular, see the shoelace entries in the *On-Line Encyclopedia of Integer Sequences* [24].

Straight-laced. Note that *straight-laced*, in the sense of 'prudish', is a modern spelling of the more correct *strait-laced*. Here 'strait', as in 'straitjacket', is an old word for 'tight' or 'narrow' and 'laced' refers to the cord used to secure a corset. Unlike 'straight-laced', the word 'strait-laced' does not suggest the presence of the horizontals in a lacing.

One-Column Lacings

Imagine pulling really hard on the two ends of the shoelace in one of your shoes that has been laced using a straight n-lacing. Then, if the lacing does not get in the way and if your foot is narrow enough, you will end up with the two rows of eyelets superimposed, one on top of the other. This means that we do not have to distinguish any longer between the two columns of eyelets and what we are dealing with is a *one-column n-lacing*, that is, a closed path in the plane consisting of n line segments whose endpoints are n equally spaced eyelets that are all contained in a vertical line. We call the one-column n-lacing that a given straight n-lacing contracts to the *contraction* of the n-lacing. For example, Figure 2.1 shows the contractions of the zigzag, star, and serpent 6-lacings. In general, there are many different straight n-lacings that contract to a given one-column n-lacing.

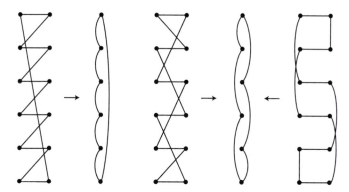

Figure 2.1. The contractions of the zigzag, star, and serpent 6-lacings.

In the following chapters, we are interested in determining the shortest and longest lacings in the different classes of n-lacings under consideration. Mainly due to the absence of diagonals, figuring out the structure of the shortest and longest one-column n-lacings is a much easier task, which we will tackle in this chapter. In particular, we derive simple algorithms for constructing all shortest and longest one-column n-lacings and calculate the lengths of these extremal one-column n-lacings and their numbers. In a final section, we describe a simple method that allows us to construct all straight n-lacings that contract to a given one-column n-lacing.

Apart from being a good warm-up exercise, our investigation of one-column n-lacings yields the results about one-column n-lacings that we will be drawing on in proofs that deal with straight n-lacings. However, in terms of its subject matter, this chapter is somewhat distinct from the rest of the book. Therefore, you may want to skip this chapter for the moment and come back to it only when you come across an argument that requires some knowledge of one-column n-lacings.

2.1 The Number of One-Column n-Lacings

It is easy to see that there is only one one-column 2-lacing, one one-column 3-lacing, and three one-column 4-lacings.

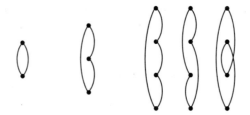

Figure 2.2. The one-column 2-, 3-, and 4-lacings.

Let's number the n eyelets 1 to n, from top to bottom. We traverse a one-column n-lacing, starting at eyelet 1, note the eyelets in the order that we come across them, and stop once we have visited all eyelets. The resulting string of numbers is a permutation of the eyelets that starts with 1. For example, the unique one-column 2-lacing corresponds to the permutation 12, and the unique one-column 3-lacing corresponds, depending on the direction in which we transverse it, to the permutation 123 or the permutation 132. In general, every one-column n-lacing, $n > 2$, corresponds to two such permutations which, in turn, correspond to the two directions in which the lacing can be transversed. Therefore, let's call a permutation of the eyelets that starts with 1 an *oriented one-column n-lacing*. There is a total of $(n-1)!$ oriented one-column n-lacings. Consequently, for $n > 2$, dividing this number by two gives the number of one-column n-lacings. We define the length and the segments of an oriented one-column n-lacing to be those of the associated one-column n-lacing.

Theorem 2.1 (Number of One-Column Lacings). *There is exactly one one-column 2-lacing, and, for $n > 2$, there are exactly $\frac{1}{2}(n-1)!$ different one-column n-lacings.*

We will exclude the trivial case $n = 2$ from all further considerations by assuming that $n > 2$ for the rest of this chapter.

In the next two sections, we determine the structure and number of the shortest and longest one-column n-lacings. The following figure shows some examples of such extremal one-column n-lacings.

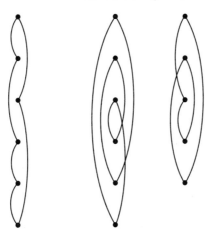

Figure 2.3. A shortest one-column 6-lacing, a longest one-column 6-lacing, and a longest one-column 5-lacing.

2.2 The Shortest One-Column n-Lacings

To simplify statements about distances as much as possible, we will fix the distance between two neighboring eyelets to be 1.

An oriented one-column n-lacing cannot be shorter than twice the distance between eyelet 1 and eyelet n, that is, $2(n-1)$. To construct an oriented one-column n-lacing of this length, we start at eyelet 1, choose a subset of the eyelets from 2 to $n-1$ and visit them in ascending order. Then we visit eyelet n and, finally, return to eyelet 1 via all the eyelets that we have not visited yet by visiting them in descending order. Clearly, every shortest oriented one-column n-lacing can be constructed like this and is therefore uniquely determined by the subset of eyelets visited on the way from eyelets 1 to n, excluding eyelets 1 and n. The number of such subsets is 2^{n-2}. Therefore, dividing this number by two gives the number of the shortest one-column n-lacings.

Theorem 2.2 (Shortest One-Column Lacings). *For $n > 2$, the number of shortest one-column n-lacings is 2^{n-3}. The length of all shortest one-column n-lacings is $2(n-1)$.*

The left diagram in Figure 2.3 shows a shortest one-column 6-lacing. Note also that it is exactly the simple-and-straight n-lacings that contract to shortest one-column n-lacings. We'll see in Section 2.4 that there are always a number of different simple-and-straight n-lacings that contract to any given shortest one-column n-lacing.

2.3 The Longest One-Column n-Lacings

Let n be an even number, and let V be a one-column n-lacing of maximal length. As we traverse V, we pick every second segment and collect all the segments picked in this way in a set V_1 and the remaining segments in a set V_2. Both V_1 and V_2 partition the set of eyelets, in the sense that every single one of the eyelets is endpoint of exactly one of the segments in V_1 and endpoint of exactly one of the segments in V_2; see Figure 2.4. We also split the set of n eyelets into two parts. The *top part* E_1 consists of the eyelets 1 to $n/2$ and the *bottom part* E_2 consists of the remaining eyelets; see, again, Figure 2.4.

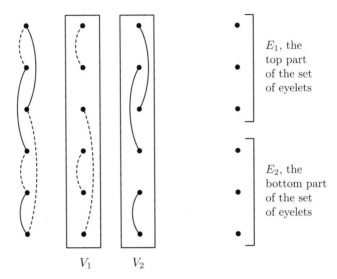

Figure 2.4. Splitting a one-column 6-lacing into the disjoint sets V_1 and V_2, and splitting the set of six eyelets into its top and bottom parts E_1 and E_2.

Let s be a segment in V_1. We want to show that the two endpoints of s are contained in different parts—one in the top and one in the bottom part. Assume that this is not the case. Then, without loss of generality, we may assume that both are contained in E_1. A simple counting argument shows that there is also at least one segment s' in V_1 whose endpoints are both contained in E_2. Then there are segments w and w' having the following properties:

- The sum of the lengths of w and w' is greater than the sum of the lengths of s and s'.
- The four endpoints of w and w' coincide with those of s and s'.
- Replacing s and s' by w and w' yields a new lacing V'.

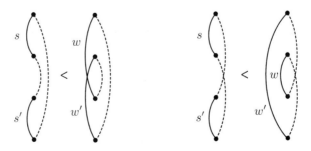

Figure 2.5. Lengthening a one-column lacing.

Figure 2.5 shows the two essentially different ways in which s and s' can be tied into the overall lacing and how w and w' have to be chosen in each case. We conclude that the new lacing V' is longer than V. This is impossible since V is of maximal length. This means that V_1 (as well as V_2) is a set of segments having the following two properties:

(P1) Every eyelet is endpoint of exactly one segment in the set.
(P2) Given one of the segments in the set, one of its endpoints is contained in the top part E_1 and the other one in the bottom part E_2.

We call a set of segments having properties P1 and P2 a *split n-partition*. The *length of a split n-partition* is just the sum of the lengths of all segments in the partition. Given a split n-partition N, we define a bijective function $E_1 \to E_2$ that maps an eyelet e in E_1 to the uniquely determined eyelet in E_2 which is connected to the eyelet e by a segment in N. Conversely, given a bijective function $E_1 \to E_2$, there is a uniquely determined split n-partition that gives rise to it as described above. We conclude that there are as many split n-partitions as there are bijective functions $E_1 \to E_2$; that is, there are exactly

$$\left(\frac{n}{2}\right)!$$

such partitions. A particularly simple example of a bijective function from E_1 to E_2 is

$$E_1 \to E_2 : e \mapsto e + \frac{n}{2}.$$

The length of the split n-partition corresponding to this function is

$$\left(\frac{n}{2}\right)^2.$$

Let P be a split n-partition and let ij and $i'j'$ be two of its segments such that $i, i' \in E_1$; see Figure 2.6. To perform a swap of eyelets i and i' on P means to replace these two segments by the segments ij' and $i'j$. By performing this swap, we construct a new split n-partition P', and it is easy to see that the length of P equals the length of P'.

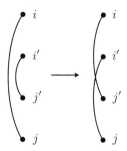

Figure 2.6.

Using successive swaps, it is possible to transform any split n-partition into any other split n-partition. This implies that the lengths of all split n-partitions are equal to $(n/2)^2$. Since a longest one-column n-lacing is made up of two split n-partitions, the length of such a lacing is

$$\frac{n^2}{2}.$$

Now, let f be the function associated with a split n-partition. Then we can construct an oriented one-column n-lacing using the following algorithm:

1. Start your journey at eyelet 1 and travel to eyelet $f(1)$;
2. If you have visited all eyelets in E_1, you have also visited all eyelets in E_2 and you complete your journey at this point by traveling back to eyelet 1.
3. Otherwise, choose one of the eyelets in E_1 that you have not visited yet, say x, and make it the next stop on your journey. Following this, travel to eyelet $f(x)$. Return to the previous step.

Clearly, every longest oriented one-column n-lacing arises like this from exactly one split n-partition in exactly one way. This means that the number of longest oriented one-column n-lacings is the number of split n-partitions times the number of different ways to construct a lacing from the associated functions. We divide this number by 2 to arrive at

$$\frac{1}{2}\left(\frac{n}{2}\right)!\left(\frac{n}{2}-1\right)!,$$

the number of longest one-column n-lacings in the case that n is an even number.

Consider Figure 2.7. It illustrates how a one-column n-lacing, n even, can be turned into a one-column $(n+1)$-lacing. Start by moving the top and bottom parts of eyelets a further 1 unit apart to make room for an additional eyelet. Split any of the segments of the lacing by introducing an additional point on it. Finally, make this point into the additional eyelet by moving it into its designated spot. We want to show that this procedure turns longest one-column n-lacings into longest one-column $(n+1)$-lacings and that all longest one-column $(n+1)$-lacings arise like this.

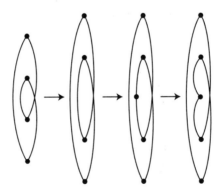

Figure 2.7. Turning a longest one-column 4-lacing into a longest one-column 5-lacing.

Let V be a longest one-column n-lacing, n odd. We split the set of eyelets into three parts: the *top part*, consisting of eyelets 1 to $(n-1)/2$; the *middle eyelet* $(n+1)/2$; and the *bottom part*, consisting of the remaining eyelets. Let c and c'

be the two segments of V ending in the middle eyelet. We want to show that the respective second endpoints of c and c' are contained in different parts; that is, one is contained in the top part and one in the bottom part. Without loss of generality, let's assume that both endpoints are contained in the top part, and let s' be one of the segments of V that has an endpoint in the bottom part. If the other endpoint of s' is also contained in the bottom part, then we can let c play the role of s in Figure 2.5 to find new segments w and w' such that replacing $s = c$ and s' in V by w and w' yields a longer one-column n-lacing. Since this is impossible, we conclude that every segment in V different from c and c' has one endpoint in the bottom part and the other in the top part. However, since both c and c' end in the top part, this is clearly impossible. This means that the endpoints of c and c' different from the middle eyelet are contained in different parts.

By reversing the procedure described above and sketched in Figure 2.7, we end up constructing a one-column $(n - 1)$-lacing V'. This new lacing is necessarily a longest one-column $(n - 1)$-lacing, because, otherwise, the one-column n-lacing constructed from a longest one-column $(n - 1)$-lacing would be a one-column n-lacing that is longer than V. We conclude that every longest one-column n-lacing, n odd, arises from a longest one-column $(n - 1)$-lacing and that every longest one-column $(n - 1)$-lacing gives rise to $n - 1$ different longest one-column n-lacings (one per segment), except in the case $n = 3$ (because both segments of the unique one-column 2-lacing correspond to the same longest 3-lacing). This means that for $n > 3$, the length of a longest one-column n-lacing equals the length of a longest one-column $(n - 1)$-lacing plus $n - 1$, that is,

$$\frac{(n-1)^2}{2} + n - 1 = \frac{n^2 - 1}{2}.$$

Furthermore, the number of different longest one-column n-lacings for $n > 3$ equals the number of different longest one-column $(n - 1)$-lacings times $n - 1$, that is,

$$\frac{n-1}{2} \left(\frac{n-3}{2} \right)! \left(\frac{n-1}{2} \right)!$$

We summarize our results about one-column n-lacings in the following theorem:

Theorem 2.3 (Longest One-Column Lacings). *For $n > 3$, there is a total of*

$$\frac{1}{2} \left(\frac{n-2}{2} \right)! \left(\frac{n}{2} \right)!$$

longest one-column n-lacings for even n, and there is a total of

$$\frac{n-1}{2} \left(\frac{n-3}{2} \right)! \left(\frac{n-1}{2} \right)!$$

longest one-column n-lacings for n odd.

The length of a longest one-column n-lacing is $n^2/2$ if n is even and $(n^2 - 1)/2$ if n is odd.

Remember that there is only one one-column 2-lacing and one one-column 3-lacing. Therefore, it does not make much sense to talk about the longest one-column 2- and 3-lacings.

2.4 Straight n-Lacings and One-Column n-Lacings

Every straight n-lacing contracts to a unique one-column n-lacing. On the other hand, every one-column n-lacing is the contraction of a number of different straight n-lacings. In the following, we demonstrate how these different straight n-lacings can be constructed.

It is easy to see that there are exactly two straight 2-lacings that contract to the unique one-column 2-lacing. One of these 2-lacings is dense, the other is superstraight; see Figure 2.8.

Figure 2.8. The two straight 2-lacings that contract to the unique one-column 2-lacing.

From now on, let again $n > 2$. We first consider an arbitrary straight n-lacing and, starting at one of the eyelets A_k, trace this lacing by moving in the direction of the horizontal that contains A_k. As we proceed, we note the different nonhorizontal segments in the order that we come across them, as well as the direction in which we transverse them (up or down). We stop as soon as we arrive back at A_k. For example, consider the left diagram in Figure 2.9. Starting at eyelet A_1, the resulting sequence of segments and directions is

4-diagonal/down,3-vertical/up,2-diagonal/down,1-vertical/up,2-diagonal/up

Our movement around the n-lacing induces, in a natural way, a movement around the contraction of this n-lacing. In the diagram the contraction is shown on the right, and both the orientation given by our original movement and the induced orientation of the contraction are indicated by arrows.

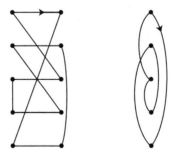

Figure 2.9. Orientation of a straight n-lacing and the induced orientation of the contraction of this n-lacing.

Clearly, the sequence of segments and directions, plus the information that we start our journey at eyelet A_k, is all that is needed to reconstruct the original straight n-lacing. We can split the sequence into three sequences: (1) 4, 3, 2, 1, 2; (2) diagonal, vertical, diagonal, vertical, diagonal; and (3) down, up, down, up, up. The first and the third sequences, together with the information that we start at A_k, is all we need to construct the one-column n-lacing that the original straight n-lacing contracts to, as well as the induced orientation of this contraction. Furthermore, the last element of the vertical/diagonal sequence is determined by the other elements of this sequence. Summarizing, this means that the original straight n-lacing is uniquely determined by the integer k, a one-column n-lacing, an orientation of this one-column n-lacing, and a vertical/diagonal sequence of length $n-1$. Conversely, it is clear that *any* choice of an integer k ($1 \leq k \leq n$), a one-column n-lacing, an orientation of this one-column n-lacing, and a vertical/diagonal sequence of length $n-1$ actually corresponds to a uniquely determined straight n-lacing. This immediately translates into a method to construct all straight n-lacings that correspond to a given one-column n-lacing.

We can now count the number of those straight n-lacings that contract to a given one-column n-lacing as follows: We may fix k to be any of the integers from 1 to n. Since there are 2^{n-1} different vertical/diagonal sequences of length $n-1$ and every one-column n-lacing has two orientations, we conclude that there are a total of 2^n different straight n-lacings that contract to any given one-column n-lacing.

Let's again start with a straight n-lacing S together with an eyelet A_k and construct the triple consisting of its contraction, the induced orientation of this contraction, and its vertical/diagonal sequence. Reversing the orientation in this triple results in a triple whose associated straight n-lacing is the vertical mirror image of the original n-lacing S. Replacing the vertical/diagonal sequence by a sequence consisting of diagonals only gives a new triple whose associated straight n-lacing is dense. Let's call this new n-lacing the *k-diagonalization of S*. Replacing the vertical/diagonal sequence by a sequence consisting of verticals only gives a new triple whose associated straight n-lacing we want to call the *k-stratification of S*. If n is even, this new straight n-lacing is superstraight. If n is odd, this lacing is a straight n-lacing whose nonhorizonal segments are all verticals except for one diagonal that contains the eyelet A_k. See Figure 2.10 for the 1-diagonalization and the 1-stratification of our original example.

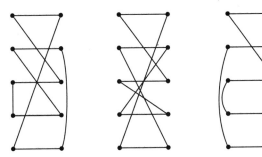

Figure 2.10. A straight n-lacing (left), its 1-diagonalization (middle), and its 1-stratification (right).

The k-diagonalizations of S are identical for all possible choices of k. It therefore makes sense to omit the k and just speak of the *diagonalization* of S. For n even, a k-stratification is either equal to the 1-stratification of S or the vertical mirror image of this 1-stratification, and we will call the 1-stratification of S the *stratification* of S.

By construction, the collection of vertical lengths of the segments in S coincides with the collections of vertical lengths of its vertical mirror image, its diagonalization, and any one of its stratifications. This property will prove very useful in Chapters 4 and 6 when it comes to deriving the shortest and longest straight and simple-and-straight n-lacings. For example, this property immediately implies that a longest straight n-lacing is dense, because the diagonalization of a straight n-lacing that is not dense is clearly longer than this straight n-lacing. On the other hand, a stratification of a straight n-lacing is often shorter than the original.

The following result is an immediate consequence of our considerations above:

Theorem 2.4 (Contractions). *There are exactly two straight 2-lacings that contract to the unique one-column 2-lacing; see Figure 2.8.*

Let $n > 2$. Then there are exactly 2^n different straight n-lacings that contract to any given one-column n-lacing. Exactly two of these straight n-lacings are dense-and-straight n-lacings, one the vertical mirror image of the other. Furthermore, if n is even, exactly two of these straight n-lacings are superstraight n-lacings, one the vertical mirror image of the other.

Finally, we collect some corollaries to some of the main results of subsequent chapters into the following neat relationship between straight n-lacings and one-column n-lacings:

Theorem 2.5 (Extreme Contracts to Extreme). *The shortest straight, dense-and-straight, simple-and-straight, dense-and-simple-and-straight, and superstraight n-lacings contract to some of the shortest one-column n-lacings.*

The longest straight, dense-and-straight, and superstraight n-lacings contract to some of the longest one-column n-lacings.

This theorem is an immediate consequence of Theorems 4.2, 6.6, and 6.7.

2.5 Notes

Open One-Column Lacings. A one-column n-lacing is a roundtrip consisting of n segments whose endpoints are the n eyelets. If, instead of roundtrips, we are interested in journeys with a beginning and an end, that is, *open* one-column n-lacings, and we are interested in the problem of finding the longest such lacings, then solutions can be found in [12], page 238, or [26], problem 64. These solutions can be easily described within our framework as follows:

In the even case, consider all those longest one-column n-lacings that contain a segment that connects the middle two eyelets; see the left pair of diagrams in Figure 2.11. Then removing these special segments gives exactly the solutions we are interested in. As an immediate consequence of Theorem 2.3, all these solutions are of length $(n^2 - 2)/2$.

In the odd case, consider all those longest one-column n-lacings that contain a segment of length 1 ending in the middle eyelet; see the right pair of diagrams in Figure 2.11. Note that both segments ending in the middle eyelet may be of length one. Removing such a segment from one of these one-column n-lacings gives one of the solutions we are interested in, and all solutions arise in this manner. This means that, by Theorem 2.3, all these solutions are of length $(n^2 - 3)/2$.

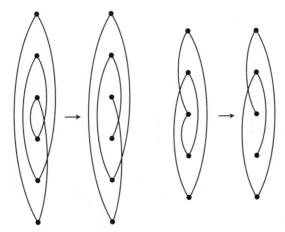

Figure 2.11. Constructing a longest open one-column 6-lacing and a longest open one-column 5-lacing.

THE BUCKETS: © United Feature Syndicate, inc.

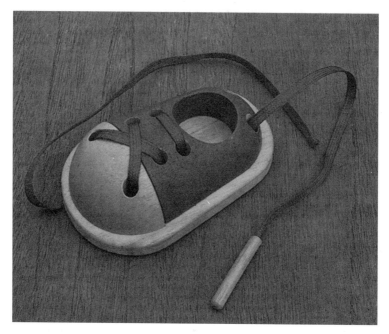

Figure 2.12. Lacing shoes are wooden toys for children to practice their lacing skills with.

3

Counting Lacings

In this chapter, we derive formulae for the numbers of n-lacings, dense n-lacings, simple n-lacings, straight n-lacings, and superstraight n-lacings, as well as the numbers of all those n-lacings that are of one of the mixed types, such as dense-and-simple, simple-and-straight, etc.

3.1 The Number of n-Lacings

It is easy to see that there are exactly three 2-lacings, and to figure out which of these three 2-lacings belong to which of the different classes under consideration; see Figure 3.1.

Figure 3.1. The 2-lacings.

The following theorem is the main result of this chapter. It gives formulae for the different numbers of n-lacings we are interested in for $n > 2$. Particularly interesting are the formula for the number of n-lacings at the top of the Venn diagram and the formula for the number of simple n-lacings at the bottom. Especially the latter is a very striking example of a simple mathematical object giving rise to a beautiful, yet surprisingly complicated, formula.

Note that the five roots of the quintic polynomial that occurs in the formula for the number of simple n-lacings are all different. Three of these roots are real and two are not. Furthermore, none of these roots is "expressible in terms of radicals". Finally, it is worth pointing out that some of the formulae in this theorem also work in the case $n = 2$.

Theorem 3.1 (Numbers of Lacings). *Let $n > 2$. Then the numbers of n-lacings of the ten different types considered by us are as follows:*

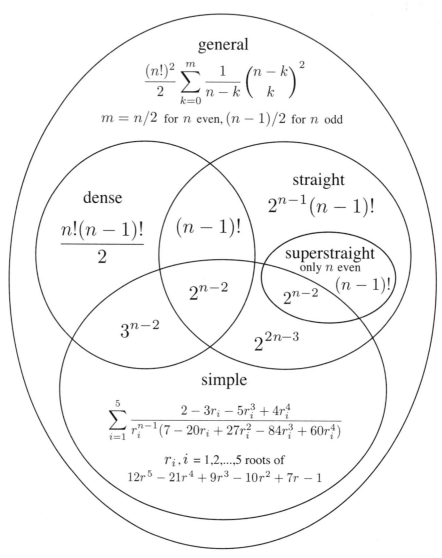

Figure 3.2. The numbers of the different kinds of n-lacings summarized in a Venn diagram.

We calculate that there are 42 different 3-lacings. These lacings are listed in Figure 3.3 up to symmetries of the underlying mathematical shoe, that is, up to reflections in the horizontal and vertical symmetry axes of this shoe, and half-turns. In particular, the diagrams in the first row correspond to one lacing each, the ones in the second and third rows to two lacings each, and the ones in the last row to four lacings each: $2 \cdot 1 + 8 \cdot 2 + 6 \cdot 4 = 42$. The six dense 3-lacings correspond to the four diagrams in the box.

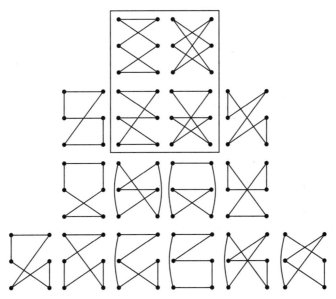

Figure 3.3. The different 3-lacings, up to symmetries of the underlying mathematical shoe.

We use Figure 3.1 and the formulae in Theorem 3.1 to determine the entries in the following table.

Table 3.1. The numbers of dense, simple, straight, and general n-lacings for small values of n.

n	2	3	4	5	6	7	8
dense	1	6	72	1440	43200	1814400	10160400
simple	2	11	57	302	1605	8521	45250
straight	2	8	48	384	3840	46080	645120
general	3	42	1080	51840	3758400	382838400	52733721600

Use the list of the simple 4-lacings in Figure 3.4, below, together with the lists of 2- and 3-lacings in Figures 3.1 and 3.3 to build up some intuition for the way lacings are built, and refer back to these lists when it comes to counting the numbers of lacings in the different classes of n-lacings for small n.

To prove the part of Theorem 3.1 that deals with general and dense n-lacings, we need to introduce *oriented n-lacings*, the algebraic counterparts of n-lacings. An oriented n-lacing l is a sequence of eyelets of length $2n$ whose elements are the $2n$ eyelets of our mathematical shoe. Furthermore, l has the following properties:

(O1) Every single one of the $2n$ eyelets is contained in l exactly once. Basically, this means that l is a permutation of the eyelets.

(O2) At most two adjacent elements of l are contained in the same column of eyelets. Here two elements are considered to be adjacent if they follow each other in the sequence or if they are the first and last elements of the sequence.

(O3) The first element of l is eyelet A_1.

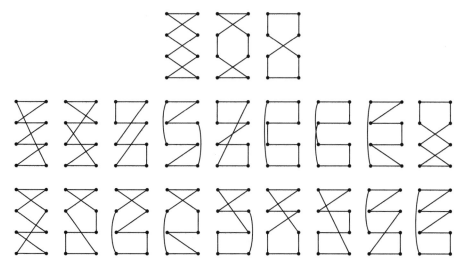

Figure 3.4. The simple 4-lacings, up to symmetries of the underlying mathematical shoe.

Given an n-lacing, we can construct an oriented n-lacing by traveling along the n-lacing and noting the eyelets in the order in which we visit them, starting and finishing our journey at eyelet A_1. There are two different directions in which an n-lacing can be transversed in this way. Corresponding to these two directions are two different oriented n-lacings. Conversely, every oriented n-lacing clearly corresponds to a uniquely determined n-lacing. We call an oriented n-lacing dense, straight, etc., if it arises from a dense, straight, etc., n-lacing, respectively.

Proof of Theorem 3.1. We deal individually with the different classes of n-lacings under consideration. Remember that $n > 2$.

Dense n-lacings. The number of dense n-lacings is the number of dense oriented n-lacings divided by two. If you forget about the indices of its entries, a dense oriented n-lacing looks as follows:

$$\underbrace{ABABAB\ldots B}_{2n}.$$

The index of the first A is always 1, and the indices of the remaining $n-1$ As and n Bs can be chosen arbitrarily as long as all possible indices are represented exactly once. We conclude that the number of dense n-lacings is $n!(n-1)!/2$.

General n-lacings. We now calculate the number of oriented n-lacings. Again, dividing this number by 2 gives the number of n-lacings that we are really after. We start with the case that n is an odd number. By Lemma 1.1, an n-lacing has an even number $2k$ of verticals, and half of these verticals are A-verticals and the other half B-verticals. Furthermore, k is one of the integers from 0 to $(n-1)/2$. Therefore, if $S(k,n)$ denotes the number of oriented n-lacings with $2k$ verticals, the number of all oriented n-lacings is

$$\sum_{k=0}^{(n-1)/2} S(k,n).$$

It remains to calculate the numbers $S(k,n)$. Consider an oriented n-lacing with $2k$ verticals. If its last element is contained in column A, then we move it to the front of the sequence. Furthermore, for the moment, forget about the indices of the elements. What we are left with is a sequence that alternates between blocks of As and blocks of Bs, one of these blocks is either a single or a double, the first block is an A-block, and there are a total of k double As, k double Bs, $n - 2k$ single As and $n - 2k$ single Bs.[1] Let $S^A(k,n)$ be the number of such sequences starting with a single A and $S^{AA}(k,n)$ be the number of such sequences starting with a double A. Then

$$S^A(k,n) = \binom{n-k-1}{k}\binom{n-k}{k},$$

and

$$S^{AA}(k,n) = \binom{n-k-1}{k-1}\binom{n-k}{k},$$

where $k = 0,1,2,\ldots,(n-1)/2$. Remember that if r is an integer and $s = -1$, then $\binom{r}{s} = 0$. In the above formula, this becomes important for evaluating the expression $S^{AA}(0,n)$.

Given one of the sequences that starts with a single A, we can reconstruct all oriented n-lacings that arises from it by adding indices to the As and Bs. This can be done arbitrarily as long as all possible indices are represented exactly once and the index of the first A is chosen to be 1. There are exactly $n!(n-1)!$ different ways of doing this.

Similarly, given one of the sequences that starts with a double A, we can reconstruct all oriented n-lacings that it arises from by adding indices to the As and Bs. This can be done arbitrarily as long as all possible indices are represented exactly once and the index of the first or the second A is chosen to be 1. There are exactly $2n!(n-1)!$ different ways of doing this.

Summarizing, we conclude that the total number of oriented n-lacings is

$$n!(n-1)! \sum_{k=0}^{(n-1)/2} \left(S^A(k,n) + 2S^{AA}(k,n) \right)$$

$$= n!(n-1)! \sum_{k=0}^{(n-1)/2} \left[\binom{n-k-1}{k}\binom{n-k}{k} + 2\binom{n-k-1}{k-1}\binom{n-k}{k} \right]$$

$$= n!(n-1)! \sum_{k=0}^{(n-1)/2} \binom{n-k}{k} \left[\binom{n-k-1}{k} + 2\binom{n-k-1}{k-1} \right]$$

$$= n!(n-1)! \sum_{k=0}^{(n-1)/2} \binom{n-k}{k} \left[\binom{n-k}{k} + \binom{n-k-1}{k-1} \right]$$

$$= n!(n-1)! \sum_{k=0}^{(n-1)/2} \binom{n-k}{k}^2 \left[1 + \frac{k}{n-k} \right]$$

---

[1] Moving a single A in the original sequence to the front ensures that, at this point, we are dealing with the same number of double As and double Bs.

$$= (n!)^2 \sum_{k=0}^{(n-1)/2} \frac{1}{n-k} \binom{n-k}{k}^2.$$

We divide by 2 to arrive at the desired formula for the number of n-lacings.

The case where n is an even number is dealt with in a similar fashion.

Simple n-lacings. There are two simple 2-lacings and eleven simple 3-lacings. These numbers can be easily extracted from the lists of lacings in Figures 3.1 and 3.3. Let $n \geq 4$.

By Lemma 1.3, every simple n-lacing "starts" with the top horizontal and "ends" with the bottom horizontal. Even stronger, it immediately follows from the definition of simple n-lacings that every such n-lacing "starts" with one of the configurations of segments listed in Figure 3.5. Let's call these configurations *fishtail, box, Zee, mirror Zee, Cee,* and *mirror Cee,* respectively.

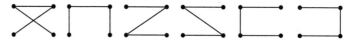

Figure 3.5. Every simple n-lacing starts with one of these configurations of segments: fishtail, box, Zee, mirror Zee, Cee, mirror Cee.

Consider a simple n-lacing that starts in a fishtail, delete this configuration, connect the resulting two loose ends by a horizontal, and what you are left with is a simple $(n-1)$-lacing. Conversely, every simple $(n-1)$-lacing can be turned into a simple n-lacing by replacing its top horizontal by a fishtail.

Consider a simple n-lacing that starts in a Zee or mirror Zee, and delete this configuration. Then you are left with two loose ends. Move the loose end that sticks out at the top one eyelet down, and connect it by a horizontal to the other loose end, and what you are left with is a simple $(n-1)$-lacing. Conversely, every simple $(n-1)$-lacing can be turned into a simple n-lacing by a Zee or a mirror Zee by performing the "inverse" of the above operation.

Consider a simple n-lacing that starts in a box, delete this configuration, connect the resulting two loose ends by a horizontal, and what you are left with is a simple $(n-1)$-lacing having the property that all segments ending in the top two eyelets are diagonals. Conversely, every simple $(n-1)$-lacing having this property can be turned into a simple n-lacing by replacing its top horizontal by a box.

This means that if we know the simple $(n-1)$-lacings, it is easy to construct and count all simple n-lacings starting in a fishtail, a box, a Zee, or a mirror Zee. To count the simple n-lacings that start with a Cee or mirror Cee is a little bit more complicated. If a simple n-lacing starts with a Cee, then it starts with one of the six configurations listed in Figure 3.6. Moving from left to right, let's call them Cee1, Cee2,..., and Cee6, respectively.

Now, let's fix some notation.

1. The *n-diagonaldiagonals* are the simple n-lacings in which the two nonhorizontal segments ending in the top two eyelets are both diagonals. Let s_n^{dd} be the number of such lacings.

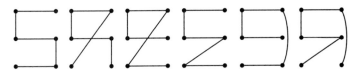

Figure 3.6. The six different Cee starting configurations.

2. The *n-verticalverticals* are the simple n-lacings in which the two nonhorizontal segments ending in the top two eyelets are both verticals. Let s_n^{vv} be the number of such lacings.

3. The *n-diagonalverticals* are the simple n-lacings in which the nonhorizontal segment ending in eyelet A_1 is a diagonal and the nonhorizontal segment ending in eyelet B_1 is a vertical. Let s_n^{dv} be the number of such lacings. Similarly, define the *n-verticaldiagonals* and their number s_n^{vd}.

4. The *Ceediagonals* are the simple n-lacings starting with a Cee in which the nonhorizontal segment ending in eyelet B_1 is a diagonal. Let s_n^{Ceed} be the number of such lacings. Similarly, define the *Ceeverticals* and their number s_n^{Ceev} as well as the *mirror Ceediagonals*, *mirror Ceeverticals*, and their respective numbers s_n^{mCeed} and s_n^{mCeev}.

It is clear that $s_n^{mCeed} = s_n^{Ceed}$, $s_n^{mCeev} = s_n^{Ceev}$, and $s_n^{dv} = s_n^{vd}$. Therefore, the total number of simple n-lacings is

$$s_n = s_n^{dd} + s_n^{vv} + 2s_n^{dv},$$

and, for fixed n, we know all the different numbers listed under 1–4 once we know $s_n^{dd}, s_n^{vv}, s_n^{dv}, s_n^{Ceed}$, and s_n^{Ceev}. Furthermore, it also makes sense to define all these numbers for $n = 2$ and $n = 3$ and extract them from Figures 3.1 and 3.3 (remember that, to start with, we did assume $n \geq 4$). In particular, for $n = 3$ we find that

$$s_3^{dd} = 4, s_3^{vv} = 3, s_3^{dv} = 2, s_3^{Ceed} = 1, s_3^{Ceev} = 1.$$

For $n = 2$ we first count

$$s_2^{dd} = 1, s_2^{vv} = 1, s_2^{dv} = 0,$$

and then, rather than adjusting the above definition, we set

$$s_2^{Ceed} = 0, s_2^{Ceev} = 0$$

to account for the special role this extreme case will play in the following. Let

$$S_n = \{s_n^{dd}, s_n^{vv}, s_n^{dv}, s_n^{Ceed}, s_n^{Ceev}\}.$$

Summarizing what we just said, we find that

$$S_2 = \{1, 1, 0, 0, 0\}, S_3 = \{4, 3, 2, 1, 1\}.$$

The following recursive system of equations can be used to calculate S_n from S_{n-1} and S_{n-2}:

$$s_n^{Ceev} = \underbrace{s_{n-2}^{dd} + s_{n-2}^{dv}}_{\text{刁}} + \underbrace{s_{n-1}^{Ceev}}_{\text{Ƨ}} + \underbrace{s_{n-2}^{dd} + s_{n-2}^{vv} + 2s_{n-2}^{dv}}_{\text{ᒪ}} + \underbrace{s_{n-2}^{dv} + s_{n-2}^{vv}}_{\text{ᔑ}}$$

$$s_n^{Ceed} = \underbrace{s_{n-2}^{dd} + s_{n-2}^{dv}}_{\text{⅄}} + \underbrace{s_{n-1}^{Ceed} + s_{n-1}^{Ceev}}_{\text{Ƨ}} + \underbrace{s_{n-1}^{Ceed}}_{\text{Ƨ}} + \underbrace{s_{n-2}^{dd} + s_{n-2}^{dv}}_{\text{ᔑ}}$$

$$s_n^{dd} = \underbrace{s_{n-1}^{dd} + s_{n-1}^{vv} + 2s_{n-1}^{dv}}_{\text{⋉}} + \underbrace{2\left(s_{n-1}^{dd} + s_{n-1}^{dv}\right)}_{\text{Z}}$$

$$s_n^{vv} = \underbrace{s_{n-1}^{dd}}_{\text{⊓}} + 2s_n^{Ceev}$$

$$s_n^{dv} = \underbrace{s_{n-1}^{dv} + s_{n-1}^{vv}}_{\text{Z}} + s_n^{Ceed}$$

The symbols under the brackets indicate where the numbers above them come from. For example, the fishtail under the first bracket in the third line means that every simple $(n-1)$-lacing, be it a diagonaldiagonal, a verticalvertical, a diagonalvertical, or verticaldiagonal, can be extended by a fishtail to a diagonaldiagonal n-lacing, as described above. Similarly, the extensions of simple $(n-1)$-lacings by boxes, Zees and mirror Zees in lines 3, 4, and 5 are those described above. The extensions indicated in the first two equations are achieved in a similar fashion. For example, the second bracket in the first line means that, given a Ceevertical $(n-1)$-lacing, we can remove the starting Cee, move up the resulting top loose end by one eyelet, and splice in the configuration under the bracket to arrive at a Ceevertical n-lacing. As a second example, the third bracket in the first line means that, given any simple $(n-2)$-lacing, we can remove the top horizontal, move up the resulting right loose end by one eyelet, and complete by the starting configuration under the bracket. In general, if the symbol under the bracket has both its ends on the right side, the upper indices of the numbers above the bracket start with a "Cee", otherwise not. To be able to restrict things to the five "variables" $s_n^{dd}, s_n^{vv}, s_n^{dv}, s_n^{Ceed}$, and s_n^{Ceev}, we have used the equalities $s_n^{mCeed} = s_n^{Ceed}$, $s_n^{mCeev} = s_n^{Ceev}$, and $s_n^{dv} = s_n^{vd}$ to substitute any s_n^{mCeed}, s_n^{mCeev}, and s_n^{vd}, by s_n^{Ceed}, s_n^{Ceev}, and s_n^{dv}, respectively. For example, the second summand in the second line should be s_{n-1}^{vd} since $(n-1)$-verticaldiagonals can, whereas $(n-1)$-diagonalverticals cannot, be extended to simple n-lacings by the configuration under the bracket.

Finally, we can use standard generating-function techniques to derive the complicated formula for the number of simple n-lacings in Figure 3.2 from this system of recursive equations and the entries of S_2 and S_3. One of the easiest ways of doing this is to use the RSOLVE package in *Mathematica*.

Dense-and-Simple n-lacings. Given a dense-and-simple n-lacing, we start at eyelet A_1 and travel down via the nonhorizontal segment ending in this eyelet. In every single one of the next $n-2$ rows we visit either 0, 1, or 2 eyelets. This gives a sequence of $n-2$ 0s, 1s, and 2s. On the other hand, every such sequence clearly gives rise in a unique way to a "way down" in a simple-and-dense n-lacing. Since in a simple-and-dense n-lacing the way down determines the way back up uniquely, this implies that the number of such lacings is 3^{n-2}.

Straight (etc.) n-lacings. The formulae for the numbers of straight, dense-and-straight, simple-and-straight, superstraight, simple-and-superstraight, and dense-and-simple-and-straight n-lacings can be easily derived from our results about one-column n-lacings. Before you read on, be sure to familiarize yourself with the terminology introduced in Section 2.4.

We collect the contractions of all the n-lacings in one of the classes under discussion into a set. Then the number of different n-lacings in the class that contracts to an element of this set is independent of which element we choose and is equal to the number of possible vertical/diagonal sequences times two (every one-column n-lacing has two orientations). This means that the number of straight n-lacings in the class we are looking at is:

$$2 \cdot (\text{no. of contractions}) \cdot (\text{no. of vertical/diagonal sequences}).$$

The following table lists for every class of n-lacings under consideration the number of one-column n-lacings that are contractions of the n-lacings in this class, and the number of possible vertical/diagonal sequences that correspond to every single one of the contractions of the n-lacings in this class.

	kind and no. of contractions	kind and no. of vertical/diagonal sequences
straight	all one-column n-lacings $\frac{1}{2}(n-1)!$	all of length $n-1$ 2^{n-1}
dense-and-straight	all one-column n-lacings $\frac{1}{2}(n-1)!$	diagonal only 1
simple-and-straight	all shortest 2^{n-3}	all of length $n-1$ 2^{n-1}
superstraight	all one-column n-lacings $\frac{1}{2}(n-1)!$	vertical only 1
simple-and-superstraight	all shortest 2^{n-3}	vertical only 1
dense-and-simple-and-straight	all shortest 2^{n-3}	diagonal only 1

In compiling this table we used: Theorem 2.1, which gives $\frac{1}{2}(n-1)!$ as the number of different one-column n-lacings; Theorem 2.2, which gives 2^{n-3} as the number of shortest one-column n-lacings; the fact that there are 2^{n-1} vertical/diagonal sequences of length $n-1$; and the fact that it is exactly the simple-and-straight n-lacings that contract to shortest one-column n-lacings. Finally, it is an easy exercise to calculate the numbers of elements in the different classes using the formula above. For example, the number of straight n-lacings is $2\frac{1}{2}(n-1)!2^{n-1} = 2^{n-1}(n-1)!$ $\square$

3.2 Notes

The Overs and Unders. Although the two top lacings in Figure 3.7 look very different, they are just two incarnations of the same serpent 4-lacing. Similarly, the photo at the bottom shows two ways of lacing a star 4-lacing. Our simple mathematical model of lacings of real shoes does not try to capture the exact manner in which a real shoelace weaves over and under itself and through the eyelets. It would be a straightforward exercise to extend our model to one that keeps track of the movement of the shoelace, to count how many different real lacings correspond to one of our mathematical lacings, and so forth.

Figure 3.7. Two incarnations of a serpent 4-lacing (top), and two of a star 4-lacing (bottom).

4

The Shortest Lacings

In [14] Halton proved that the crisscross n-lacing is the shortest among all those dense n-lacings that contain the horizontal segment connecting the top pair of eyelets.[1] Later, Misiurewicz [20] gave a very short proof of this result. In fact, he showed that this result stays true even if the eyelets are not fully aligned, as in our set-up. In [16] Isaksen sketched a proof of the fact that the bowtie n-lacing(s) are the shortest among all those n-lacings that contain the horizontal segment connecting the top pair of eyelets.

4.1 Statement of Results

In this chapter, we first generalize Halton's and Isaksen's results.

Theorem 4.1 (Shortest General and Dense). *The bowtie n-lacings are the shortest n-lacings overall, and the crisscross n-lacing is the shortest dense n-lacing.*

We give a proof of this result in Section 4.2 and a sketch of another proof on page 45. Then, in Section 4.3, we prove the following theorem:

Theorem 4.2 (Shortest Straight). *If n is even, then the simple-and-superstraight n-lacings are the shortest straight n-lacings. If n is odd, then the zigsag n-lacings are the shortest straight n-lacings.*

This result implies that the serpent n-lacings are among the shortest straight n-lacings. In Section 4.4 we prove the following result:

Theorem 4.3 (Shortest Dense-and-Straight). *The star n-lacings are the shortest dense-and-straight n-lacings.*

---

[1] Note that it is along the top horizontal that most of us tie our shoes. However, there are certain riding boots, such as the *Dehner* boots with nine pairs of eyelets (see [5]), that are tied along a different horizontal. Also, remember that by Lemma 1.3 every simple n-lacing contains the top and bottom horizontals.

Together, these three results imply that all shortest n-lacings in the different classes of n-lacings under consideration are simple. Therefore, we can summarize our theorems visually as in Figure 4.1. Note that the types of the shortest lacings in the different classes do not change as we vary the stretch of our mathematical shoe.

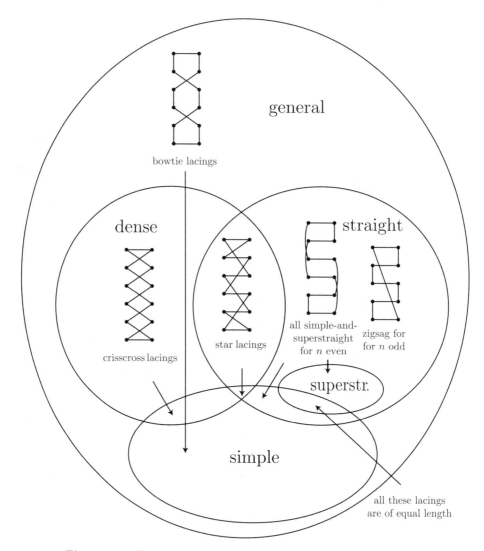

Figure 4.1. The shortest lacings in the different classes of n-lacings.

Remember that the bowtie and serpent 2-lacings coincide and that the crisscross, zigzag, and star 2-lacings coincide. Now, Theorems 4.1, 4.2, and 4.3 immediately imply the first part of the following theorem. We prove the second part in Section 4.5.

Theorem 4.4 (Relative Lengths of Popular Lacings). *Lengthwise the criss-cross, zigzag, star, serpent, zigsag, and bowtie n-lacings compare as follows:*

$$n = 2 : |bowtie| = |serpent| < |crisscross| = |star| = |zigzag|$$

$$n > 2 : |bowtie| < |serpent| \begin{array}{c} |crisscross| \\ (n \text{ even}) \\ |zigsag| \ (n \text{ odd}) \end{array} < |star| < |zigzag|$$

If $n > 2$ and even, then there is a positive number h_{sn} such that the following holds: If the stretch of our mathematical shoe is greater than, equal to, or less than h_{sn}, then the length of the crisscross n-lacing is less than, equal to, or greater than the length of the serpent n-lacings, respectively.

Similarly, if $n > 2$ and odd, then there is a positive number h_{zn} such that the following holds: If the stretch of our mathematical shoe is greater than, equal to, or less than h_{zn}, then the length of the crisscross n-lacing is less than, equal to, or greater than the length of the zigsag n-lacings, respectively.

Finally, $\lim_{n \to \infty} h_{sn} = \lim_{n \to \infty} h_{zn} = \frac{3}{4}$.

We note that the relative lengths of the crisscross, star, and zigzag n-lacings were already determined by Halton [14].

4.2 The Shortest General and Dense n-Lacings

Here is an outline of our strategy for proving Theorem 4.1, which asserts that the shortest n-lacings are the bowtie n-lacings and the shortest dense n-lacings are the crisscross n-lacings. We first define a set of "local shortening rules". These rules allow us to modify small parts of most n-lacings to come up with shorter n-lacings. Then it is easy to describe the *reduced* n-lacings, that is, the n-lacings to which none of the shortening rules apply, and to show that the bowtie n-lacings are the shortest among the reduced n-lacings.

The shortening rules also have the property that when applied to an n-lacing, they produce an n-lacing that has the same number of verticals as the one we started with. This implies that, applied to a dense lacing, the shortening rules produce a shorter dense lacing. It turns out that the only reduced dense n-lacing is the crisscross n-lacing. This immediately identifies the crisscross n-lacing as the shortest dense n-lacing.

Proof of Theorem 4.1. A *zigzag* is a connected part of a lacing consisting of two or three segments of the form diagonal-diagonal or diagonal-vertical-diagonal. The two eyelets which form the endpoints of the zigzag are called the *feet* of the zigzag and the remaining eyelet(s) its *horn(s)*. The *head* of a zigzag consists of its horn(s) and, if it has two horns, the segment connecting the horns. Figure 4.2 shows the three essentially different kinds of zigzags.

Figure 4.2. The three different zigzags.

It is clear that a dense lacing contains only zigzags of the first kind. We claim that a shortest n-lacing does not contain any of the twisted zigzags depicted on the right. To see this, consider this diagram not as consisting of three straight line segments but rather as the union of five straight line segments a, b, c, d, and e, as indicated in Figure 4.3 on the left. Now, together, segments a and c connect the foot A with the horn C. However, the straight line segment connecting A with C is shorter than a and c together. Similarly, the straight line segment connecting B with D is shorter than b and d. This means that, if we find a zigzag like the one on the left in a lacing, we can replace it by the zigzag on the right to arrive at a shorter lacing. This is our first shortening rule. It is summarized by the diagram in the middle, which indicates how the line segments are recombined. If you then imagine anything going from left to right as being made from rubber bands, you can immediately see things snap into place into the shorter zigzag on the right. We will use this kind of graphical abbreviation for all our shortening rules.

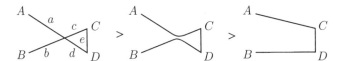

Figure 4.3. Shortening a twisted zigzag.

Figure 4.4 contains a pictorial proof that the bowtie 2-lacing is shorter than the two other 2-lacings.

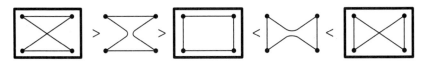

Figure 4.4. Pictorial proof that the bowtie 2-lacing is the shortest 2-lacing.

Since the crisscross 2-lacing is the only dense 2-lacing, it is also the shortest dense 2-lacing. We conclude that the theorem is true in the case $n = 2$. In the following, we will therefore assume that $n > 2$.

Now, to define the other shortening rules, we consider all possible configurations of two zigzags of the first and second kinds whose feet are on the same column. Figure 4.5 shows a typical example of a shortening rule. Note that in this example we are dealing with four "rubber bands".

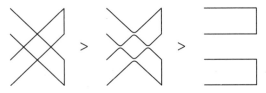

Figure 4.5. A typical shortening rule.

Using our fundamental recombining-and-straightening idea, it is usually not difficult to derive the shortening rule or rules for any conceivable reducible configuration. Figure 4.6 lists all possible shortening rules involving two zigzags of the first kind. Figure 4.7 shows the remaining shortening rules.

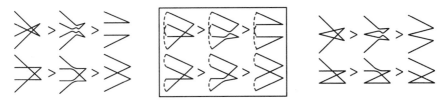

Figure 4.6. The shortening rules involving pairs of zigzags of the first kind, up to reflections in vertical and horizontal axes.

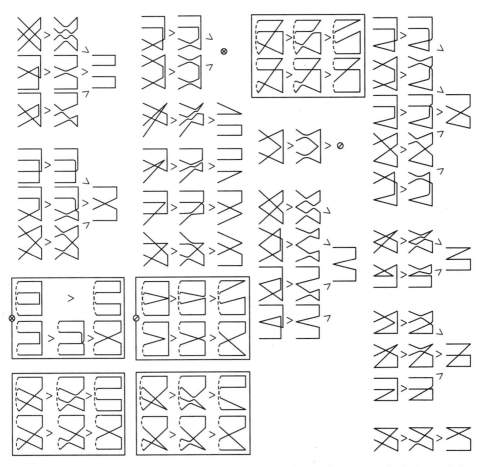

Figure 4.7. The shortening rules involving two zigzags, at least one of which is of the second kind, up to reflections in vertical and horizontal axes.

There is one minor complication that we have not discussed yet. This is illustrated by the two boxed rules in Figure 4.6. Note that these two rules apply to the same configuration. Here, the dashed curves indicate how the configuration fits into the overall lacing. Depending on which of the two possibilities we are dealing with, one or the other shortening rule has to be applied to guarantee that, after applying the rule, we are still dealing with just one loop and not two disjoint loops. In general, we get one or two shortening rules for every possible configuration of zigzags.

Those configurations that correspond to the left-hand sides of shortening rules are called *reducible* and those that correspond to right-hand sides are called *reduced*. Collecting all the reduced configurations in Figures 4.6 and 4.7, we arrive at the complete list of reduced configurations in Figure 4.8.

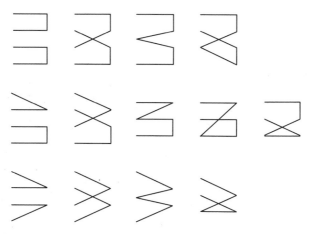

Figure 4.8. The reduced configurations of zigzags for $n > 2$, up to reflections in vertical and horizontal axes.

We inspect our list of reduced configurations of zigzags and make the following important observations:

(R1) The heads of the two zigzags in a reduced configuration never overlap (it should be clear what this means).

(R2) Given a diagonal d in a reduced lacing and a zigzag that does not contain this diagonal, at most one of the diagonals in the zigzag can intersect d in a point that is not an eyelet.

Property R1 immediately implies the following:

(R3) A vertical in a reduced lacing is a 1-vertical.

Figure 4.9 shows three reduced 5-lacings (a fairly generic one, the crisscross 5-lacing, and a bowtie 5-lacing) and one reducible lacing, which is the zigzag 5-lacing. It is easy to check that the crisscross and bowtie lacings are reduced and that the zigzag n-lacings are reducible.

Now, we need to figure out what exactly a reduced lacing looks like in general; see Figure 4.10. We start by showing that a reduced lacing contains the top horizontal $A_1 B_1$. Let's assume that this horizontal is not contained in the lacing and

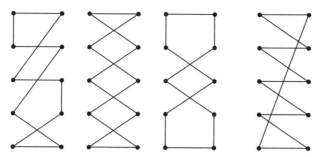

Figure 4.9. Three reduced and one reducible 5-lacing.

let's consider a diagonal d ending in eyelet A_1 and a diagonal d' ending in eyelet B_1. The second endpoint of d cannot be connected to B_1 by a vertical, since this vertical together with the two diagonals would form a zigzag of the third kind, the kind of zigzag that cannot occur in a reduced n-lacing. Hence, there is a zigzag z containing d' whose feet are contained in column A and none of whose feet and horns are endpoints of d. Then it is easy to convince ourselves, using property R1 (or R3) above, that both diagonals of z intersect the diagonal d in two points that are not eyelets, as shown in Figure 4.10. This contradicts property R2. We conclude that the top horizontal is contained in all reduced lacings.

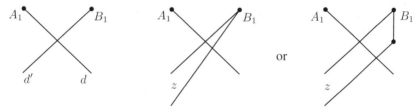

Figure 4.10. If the top end of a reduced lacing did not contain the top horizontal, it would look like one of the two diagrams on the right, which is impossible.

Starting at eyelets A_1 and B_1, we now trace the reduced lacing down along columns A and B to the first eyelets A_k and B_l, respectively, that are endpoints of diagonals different from the top horizontal. Because of property R3, k equals 1 or 2, and so does l. This means that the top end of our lacing looks like one of the diagrams in Figure 4.11.

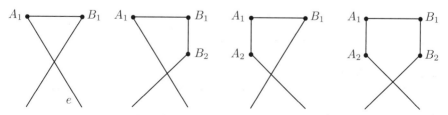

Figure 4.11. The top end of a reduced lacings looks like one of these four diagrams.

Let's consider the case shown on the left. If the diagonal e does not end in eyelet B_2, then there is a zigzag with feet in column A, one of whose horns is B_2 and none of whose horns is an endpoint of e. This implies that both diagonals of this zigzag would intersect the diagonal e in two points that are not eyelets, which, by property R2, is impossible. We conclude that e ends in eyelet B_2. We use the same argument to show that the top end of any reduced lacing looks like one of the diagrams in Figure 4.12. In other words, a new diagonal that starts at A_k ends at B_{l+1}, and a new diagonal that starts at B_l ends at A_{k+1}.

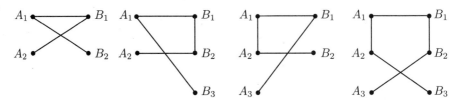

Figure 4.12. The top end of a reduced lacing looks like one of these four diagrams.

From now on things repeat. From eyelets A_{k+1} and B_{l+1} we trace the lacing down along columns A and B to the first eyelets $A_{k'}$ and $B_{l'}$, respectively, that are endpoints of diagonals different from those that we already encountered. If the two diagonals coincide, we have finished constructing the lacing, and the diagonal has to be the bottom horizontal. Otherwise, we conclude, as above, that the new diagonal starting at $A_{k'}$ has to end at $B_{l'+1}$ and that the new diagonal starting at $B_{l'}$ has to end at $B_{k'+1}$.

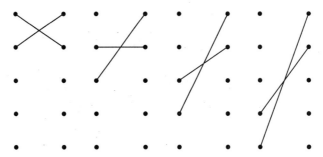

Figure 4.13. The different general crosses, up to positioning in the underlying mathematical shoe.

We define a *general cross* to consist of two intersecting diagonals whose endpoints on both columns A and B are adjacent. The *length* of a general cross is the sum of the lengths of its two diagonals. Note that a cross, as we defined it to describe bowtie lacings, is just a special general cross. The general cross on the far left of Figure 4.13 is a cross. From our discussion above, we conclude that

(R4) a reduced n-lacing is made up of the bottom and top horizontals, a number $2k$ of 1-verticals (k may be 0), and $n - 1 - k$ general crosses.

We want to show that a cross is shorter than any other general cross. This is clear for all general crosses except for those consisting of a horizontal and a 2-diagonal; see the second diagram in Figure 4.13. However, it is also an easy exercise to show that, for all $h > 0$, the length $2\sqrt{1 + h^2}$ of a cross is less than the length $1 + \sqrt{1 + 4h^2}$ of one of these special general crosses.

Furthermore, it is clear that the sum of two 1-verticals is shorter than the length of any general cross. We conclude that if we can construct a reduced lacing that contains the top and bottom horizontals, the maximum possible number of 1-verticals, and all of whose general crosses are crosses, then this lacing is of minimal length and every lacing of minimal length has to be of this type. Of course, what we have described here are exactly the bowtie n-lacings. This shows that the shortest n-lacings are the bowtie n-lacings.

It is also clear that the only reduced dense n-lacing is the crisscross n-lacing. Since applying one of our shortening rules to a dense lacing yields a dense lacing, this immediately implies that the crisscross n-lacing is the shortest dense n-lacing. This concludes the proof of Theorem 4.1.   □

4.3 The Shortest Straight n-Lacings

This section contains our proof of Theorem 4.2, which says that, for even n, the simple-and-superstraight n-lacings are the shortest straight n-lacings. Furthermore, if n is odd, then the zigsag n-lacings are the shortest straight n-lacings.

Proof of Theorem 4.2. We first show that, for even n, the shortest straight n-lacings are the superstraight n-lacings that contract to the shortest one-column n-lacings or, equivalently, the simple-and-superstraight n-lacings.

Let n be even and let l be a straight n-lacing. Then the stratification of l (see Section 2.4) is a superstraight n-lacing for which the collection of vertical lengths of its segments coincides with the collection of vertical lengths of the segments of l. This implies that if l contains diagonals that are not horizontals, then the stratification of l is shorter than l. We conclude that the shortest straight n-lacings are superstraight n-lacings. Finally, since the length of a superstraight n-lacing is n plus the length of its contraction, it follows that the shortest straight n-lacings are the superstraight n-lacings that contract to the shortest one-column n-lacings. This finishes the proof of the first part of this theorem.

Next, we want to show that for n odd the shortest straight n-lacings are the zigsag n-lacings. Let l be a straight n-lacing. As a consequence of Lemma 1.1, the lacing l contains at least one diagonal that is not a horizontal. Let A_k be an eyelet contained in such a diagonal. If l contains a second such diagonal, then we can conclude as above that the k-stratification of l (see Section 2.4) is a straight n-lacing that is shorter than l. This implies that a shortest straight n-lacing s contains exactly one diagonal that is not a horizontal.

Now, whatever vertical length the diagonal in s is, we can fit this diagonal into a simple-and-straight n-lacing s' that, except for this diagonal, consists of horizontals and verticals only. This immediately implies that s has to be simple itself. Figure 4.14 illustrates how s' can be built. Let's assume that the diagonal under discussion is a k-diagonal. Then the length of s is

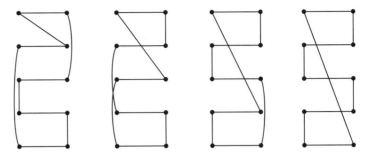

Figure 4.14. Examples of simple-and-straight 5-lacings with a 1-, 2-, 3-, 4-diagonal, respectively, as their only segment that is neither horizontal nor vertical.

$$n + (n-1)h + (n-1-k)h + \sqrt{1 + (kh)^2}.$$

We want to show that a simple-and-straight n-lacing that contains an $(n-1)$-diagonal, that is, a diagonal of maximal length, is shorter than any simple-and-straight n-lacing that contains a k-diagonal with $1 \leq k < n-1$. This means we have to show that

$$\sqrt{1 + ((n-1)h)^2} < (n-k-1)h + \sqrt{1 + (kh)^2},$$

for $1 \leq k < n-1$. However, it is very easy to see that this is the case, because the left side of this inequality corresponds to the distance between the eyelets A_1 and B_n (see Figure 4.15), whereas the right side corresponds to the length of a path connecting the same two eyelets that is not a straight line (the dotted path).

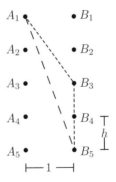

Figure 4.15.

Finally, it is very easy to see that the only simple-and-straight n-lacings that contain an $(n-1)$-diagonal are the zigsag n-lacings. This proves that the zigsag n-lacings are the shortest straight n-lacings for odd n and completes the proof of Theorem 4.2. □

4.4 The Shortest Dense-and-Straight n-Lacings

This section contains our proof of Theorem 4.3, which asserts that the star n-lacings are the shortest dense-and-straight n-lacings. We first show that the proof of this result can be reduced to proving Lemma 4.5, below.

The theorem is true for $n = 2$. In the following let $n > 2$. The collection V of the vertical lengths of the nonhorizontal segments in a straight n-lacing has the following properties:

(V1) V consists of n positive integers less than or equal to $n - 1$.
(V2) The sum of the integers in V is greater than or equal to $2(n - 1)$.

Here property V1 follows from the fact that every straight n-lacing contains n segments that are not horizontals and that the vertical length of any segment in an n-lacing is less than or equal to $n - 1$. Starting at eyelet A_1, we travel along the lacing to eyelet A_n and back. In the course of this journey, we cover a vertical distance that is greater than or equal to $2(n - 1)$. This implies property V2.

We call a set V that has properties V1 and V2 a *soft straight n-lacing*. Note that not every soft straight n-lacing necessarily arises from a straight n-lacing. Moreover, if a soft straight n-lacing does arise from a straight n-lacing, then it will, in general, arise from a number of very different straight n-lacings.

We define the length of a soft straight n-lacing V to be the sum

$$\sum_{s \in V} \sqrt{1 + (hs)^2},$$

where h is a positive number (the stretch of our shoe). Clearly, the length of a soft straight n-lacing that arises from a *dense* straight n-lacing is the same as the length of the n-lacing minus n, the sum of the lengths of the n horizontals.

The *soft star n-lacing* is the soft straight n-lacing that arises from the star n-lacings. It consists of two 1s and $n - 2$ 2s. This soft straight n-lacing turns out to be very special.

Lemma 4.5 (Star Lacing). *Let $n > 2$. Then*

1. *The only dense-and-straight n-lacings that give rise to the soft star n-lacing are the two star n-lacings.*
2. *The soft star n-lacing is the only soft straight n-lacing such that all its elements are 1s and 2s and the sum of its elements is $2(n - 1)$.*
3. *The soft star n-lacing is the shortest soft straight n-lacing.*

Clearly, together Lemma 4.5.1 and 4.5.3 imply that the star n-lacings are the shortest dense straight n-lacings.

Proof of Lemma 4.5. To prove the first part of this lemma, we need to find out what happens when we try to build up an n-lacing from n horizontals, 2 1-diagonals, and $n - 2$ 2-diagonals. To start with, it is clear that all the horizontals are present as in the left diagram of Figure 4.16.

If the nonhorizontal diagonals ending in eyelets A_1 and B_1 were both 1-diagonals or both 2-diagonals, then this would create a closed path prematurely. This means that one of these diagonals has to be a 1-diagonal and the other a 2-diagonal. Let's assume that the 1-diagonal ends in A_1, as in the right diagram of Figure 4.16. If

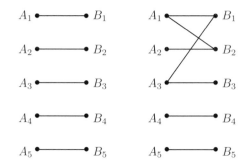

Figure 4.16. Building a dense straight n-lacing from 1- and 2-diagonals.

the nonhorizontal diagonal ending in A_2 is a 1-diagonal, we also immediately get a closed path. This is okay if we are trying to construct a 3-lacing, in which case the resulting n-lacing is a star 3-lacing, as claimed. Otherwise, this closed path comes too early, and we see that the nonhorizontal diagonal containing A_2 is necessarily a 2-diagonal. Repeating this argument a number of times yields that any dense straight n-lacing built from 1- and 2-diagonals is necessarily a star n-lacing. This concludes the proof of the first part of our lemma.

If the sum of the elements of a soft straight n-lacing is $2(n-1)$ and all its elements are 1s and 2s, then

$$k \cdot 1 + (n-k) \cdot 2 = 2(n-1),$$

for some integer k. However, solving this equation for k yields that $k = 2$ and that, therefore, the soft straight n-lacing under discussion is the soft star n-lacing. This proves the second part of the lemma.

Since, by assumption, $n > 2$, it follows that $n < 2(n-1)$. Hence,

(V3) every soft straight n-lacing contains elements that are greater than 1.

If V is a soft straight n-lacing the sum of whose elements is greater than $2(n-1)$, then we may replace one of its elements e that is greater than 1 by $e-1$ to arrive at a shorter soft straight n-lacing. This means that the sum of the elements of a shortest soft straight n-lacing is $2(n-1)$.

Let W be a soft n-lacing whose elements add up to $2(n-1)$. Then W contains a 1. Let's assume that W has an element $e > 2$. Then we replace one of its 1s by 2 and an e by an $e-1$ to arrive at a new soft straight n-lacing W' whose elements also add up to $2(n-1)$. We want to show that the length of W' is less than that of W, which immediately implies that the soft star n-lacing is the shortest soft straight n-lacing. For this it suffices to prove that

$$\sqrt{1 + (1h)^2} + \sqrt{1 + (eh)^2} > \sqrt{1 + (2h)^2} + \sqrt{1 + ((e-1)h)^2}.$$

Here is a nice geometrical argument that does just that. In Figure 4.17 the first sum corresponds to the length $|F_1 S_1 F_2|$ of the dashed path $F_1 S_1 F_2$ and the second sum to the length $|F_1 S_2 F_2|$ of the dotted path $F_1 S_2 F_2$. The two confocal ellipses E_1 and E_2 with common foci F_1 and F_2 through the points S_1 and S_2, respectively, are disjoint by construction and E_2 is contained in E_1. Now it is clear that

$$|F_1 S_1 F_2| = |F_1 T_1 F_2| > |F_1 T_2 F_2| = |F_1 S_2 F_2|.$$

This completes the proof of the lemma.   □

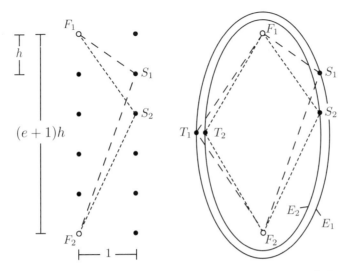

Figure 4.17. The path $F_1 S_1 F_2$ is longer than the path $F_1 S_2 F_2$.

4.5 Comparing the Lengths of Some Popular n-Lacings

This section contains our proof of Theorem 4.4. The first part of this theorem is a corollary of Theorems 4.1, 4.2, and 4.3. Here is what we still need to prove.

If $n > 2$ and even, then there is a positive number h_{sn} such that the following holds: if the stretch of our mathematical shoe is greater than, equal to, or less than h_{sn}, then the length of the crisscross n-lacing is less than, equal to, or greater than the length of the serpent n-lacings, respectively.

Similarly, if $n > 2$ and odd, then there is a positive number h_{zn} such that the following holds: if the stretch of our mathematical shoe is greater than, equal to, or less than h_{zn}, then the length of the crisscross n-lacing is less than, equal to, or greater than the length of the zigsag n-lacings, respectively.

Finally, $\lim_{n\to\infty} h_{sn} = \lim_{n\to\infty} h_{zn} = \frac{3}{4}$.

Proof of Theorem 4.4. Let $n > 2$ be an even number. Then we first want to compare the lengths of the crisscross and the serpent n-lacings; that is, we need to figure out for what possible choices of h the circle in the following line stands for $>$, $<$, or equality.

$$|\text{crisscross}| = 2 + 2(n-1)\sqrt{1+h^2} \; \bigcirc \; 2(n-1)h + n = |\text{serpent}|.$$

We conclude that

$$\sqrt{1+h^2} - h \; \bigcirc \; \frac{n-2}{2(n-1)}.$$

The right side is a number between 0 and 1/2. The left side, considered as a function $[0, \infty[\to \mathbf{R}$ in h, is a strictly decreasing continuous function which takes on the value 1 at $h = 0$ and tends to 0 as h goes to infinity. By the intermediate value theorem, there will be a unique $h_{sn} > 0$ such that the circle will be a $>$ sign for $h < h_{sn}$, will stand for equality for $h = h_{sn}$, and will stand for a $<$ sign for $h > h_{sn}$, as claimed in the theorem.

Let $n > 2$ be an odd number. Then we want to compare the lengths of the crisscross and the zigsag n-lacings. More precisely, we want to show that there is a unique $h_{zn} > 0$ such that the circle in

$$|\text{crisscross}| = 2 + 2(n-1)\sqrt{1+h^2} \bigcirc n + (n-1)h + \sqrt{1 + ((n-1)h)^2} = |\text{zigsag}|$$

will be a $>$ sign for $h < h_{zn}$, will stand for equality for $h = h_{zn}$, and will stand for a $<$ sign for $h > h_{zn}$.

Subtracting the left side from the right side, we arrive at the function

$$z_n : \mathbf{R} \to \mathbf{R} : h \mapsto n - 2 + (n-1)h + \sqrt{1 + (n-1)^2 h^2} - 2(n-1)\sqrt{1+h^2}.$$

Then we can finish our proof via the intermediate value theorem by showing that this function is negative for $h = 0$, is positive for large h, and has positive derivative for all $h > 0$.

Indeed, for $h = 0$ the function z_n takes on the value $-(n-1)$, which is negative, and for large h it gets approximated by the function

$$\mathbf{R} \to \mathbf{R} : h \mapsto n - 2 + (n-1)h + (n-1)h - 2(n-1)h = n - 2.$$

This means z_n is positive for large h. It remains to show that the derivative of z_n,

$$z_n' : \mathbf{R} \to \mathbf{R} : h \mapsto (n-1)\left(1 - \frac{2h}{\sqrt{1+h^2}} + \frac{h(n-1)}{\sqrt{1+h^2(n-1)^2}}\right),$$

is positive for $h > 0$.

This is the case if and only if the function

$$\mathbf{R} \to \mathbf{R} : h \mapsto 1 - \frac{2}{\sqrt{\frac{1}{h^2}+1}} + \frac{1}{\sqrt{\frac{1}{h^2(n-1)^2}+1}}$$

is positive for $h > 0$. But this follows immediately from the following inequalities:

$$\frac{1}{\sqrt{\frac{1}{h^2}+1}} < \frac{1}{\sqrt{\frac{1}{h^2(n-1)^2}+1}} < 1.$$

To finish the proof of this theorem, we calculate $\lim_{n \to \infty} h_{sn}$ and $\lim_{n \to \infty} h_{zn}$. From our considerations above it follows that $\lim_{n \to \infty} h_{sn}$ is the solution of the equation

$$0 = \sqrt{1+h^2} - h - \lim_{n \to \infty} \frac{n-2}{2(n-1)} = \sqrt{1+h^2} - h - \frac{1}{2},$$

which is $\frac{3}{4}$. Since h_{zn} is the positive solution of the equation $z_n(h) = 0$, it is also the positive solution of the equation

$$0 = \frac{z_n(h)}{n} = 1 - \frac{2}{n} + \frac{n-1}{n}h + \sqrt{\frac{1}{n^2} + \left(\frac{n-1}{n}\right)^2 h^2} - 2\frac{n-1}{n}\sqrt{1+h^2}.$$

Therefore, $\lim_{n \to \infty} h_{zn}$ is the solution of the equation

$$0 = 1 + h + h - 2\sqrt{1+h^2} = 1 + 2h - 2\sqrt{1+h^2}.$$

We conclude that $\lim_{n \to \infty} h_{zn} = \frac{3}{4}$. $\square$

4.6 Notes

The spectra of n-lacings. It is interesting to order the n-lacings according to length and to see how this order changes as we change the stretch of our mathematical shoe. Figure 4.18 shows the orders of the 2-lacings and 3-lacings for very short and very long shoes. As expected, the absolute shortest lacings stay the same. On the other hand, the absolute longest lacings change. We determine the longest n-lacings in Chapter 6. Because 1-verticals are the shortest segments that fit into a short shoe, we expect to see for very short shoes that the first couple of entries in such an ordered sequence of lacings are lacings that contain the maximum possible number of 1-verticals. As you can see, this is exactly what happens in the case of the 2- and 3-lacings.

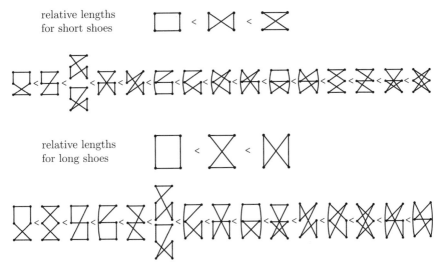

Figure 4.18. The 2-lacings and the 3-lacings (up to symmetries of the underlying mathematical shoe) ordered by length for very short shoes above and very long shoes below.

Another interesting feature of the diagrams are the two stacked 3-lacings. On closer inspection, it turns out that both contain the same kind and number of segments—one horizontal, two 1-verticals, two 1-diagonals, and one 2-diagonal. This suggests designing some puzzles based on lacings: Given a number of building blocks, how many lacings can you build?

The spectra of lacings also suggest the following 'application': Say you buy a pair of shoes that are laced zigzag. After a while one of the shoelaces breaks. Now, instead of throwing the two pieces of shoelace away, why not check whether one of them is still long enough to lace the shoe crisscross or bowtie? Of course, the other lacings in a spectrum allow for many intermediate lengths. One of the classical n-lacings, the crisscross n-lacing, turns out to be the shortest dense n-lacing. In Chapter 6, we prove the surprising result that the other classical n-lacings, the zigzag n-lacings, are the longest simple n-lacings. So, if you use dense-and-simple n-lacings to lace your shoes and are keen on this application, you should start with a zigzag n-lacing and work your way down to the crisscross n-lacing.

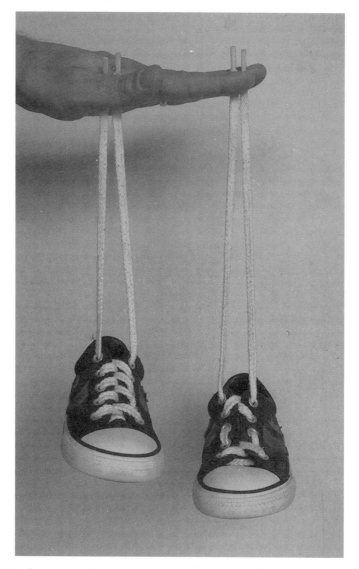

Figure 4.19. Comparing the lengths of the crisscross and bowtie lacings of a real shoe.

Shortest shoelaces in other classes of n-lacings. The shortening rules that allowed us to shorten general n-lacings all leave the number of verticals unchanged. Furthermore, the remaining arguments that we used to conclude that the bowtie n-lacings are the shortest n-lacings can also be used to conclude that, more generally, the shortest among the n-lacings that contain a fixed number $2k$ of verticals are exactly the n-lacings that contain the top and bottom horizontals, k gaps, and $n - k - 1$ crosses. The bowtie n-lacings then result when we set k to the maximal possible value, and the crisscross n-lacings result when we set k to the minimal possible value $k = 0$.

Lowest bounds. We have seen that the structure of the shortest n-lacings is independent of the stretch h of the underlying mathematical shoe. However, the total length of the shortest n-lacings does depend on h. It is clear that the smaller we choose h, the smaller this length will be. This means that if $l(h)$ denotes this length, then a lower bound for the length of an n-lacing is $\lim_{h \to 0} l(h)$, which is just the number of diagonals in the shortest n-lacings (times the horizontal separation 1). This means that this lower bound, which will never be attained by any n-lacing of any mathematical shoe, is $n + 1$ for n odd and n for n even. The corresponding lower bound for the lengths of dense n-lacings is $2n$, the number of diagonals in a dense n-lacing.

Proofs. The key idea underlying many proofs in this book is to investigate collections of numbers that share some of the properties of the collections of vertical lengths of segments in n-lacings. By using a number of simple shortening or strengthening rules, it is usually quite easy to identify those among these collections that are optimal in terms of length or strength. Since we can also show that these optimal collections arise from certain n-lacings, we conclude that these n-lacings are the optimal n-lacings.

Note that we proved Theorem 4.3 using this approach. On the other hand, in the proof of Theorem 4.1, we applied shortening rules to the n-lacings themselves. This "higher-level" approach can also be easily modified to apply in some of the more general shoelacing settings that we will be considering in the next chapter. However, the resulting proof is a little bit longer than is absolutely necessary. A shorter proof of Theorem 4.1 results when we consider collections of numbers that share some of the properties of collections of the vertical lengths of segments in n-lacings. Here is a sketch of such a proof.

Sketch of a second proof of Theorem 4.1. Let's consider collections of $2n$ pairs (x, y), where x is either the letter 'd' (for 'diagonal') or the letter 'v' (for 'vertical') and y is a nonnegative integer. The *vertical length* of such a pair is its second component, and the vertical length of a collection of such pairs is the sum of the vertical lengths of its pairs. Let's call such a collection a *perfect n-collection* if it has the following properties:

(P1) If a pair has a 0 as its second component, then its first component is a d. There are at most n pairs that have a 0 as their second component.

(P2) If n is odd or even, there are at most $n - 1$ or n pairs, respectively, that have a v as their first component.

(P3) The vertical length of the collection is greater than or equal to $2(n - 1)$.

Recalling that h is the stretch of our mathematical shoe, we define the *length* of a pair (v, y) to be hy and that of a pair (d, y) to be $\sqrt{1 + (hy)^2}$. Furthermore, we define the length of a perfect n-collection to be the sum of the lengths of its pairs. It should be clear how an n-lacing can be turned into a perfect n-collection and that the length of an n-lacing and its associated perfect n-collection are equal.

Here are five shortening rules that allow us to turn most perfect n-collections into shorter perfect n-collections:

(S1) If the vertical length of the collection is greater than $2(n - 1)$, replace one pair (x, y) by $(x, y - 1)$, ensuring that condition P1 is still satisfied.

(S2) If there are less than the maximal number of pairs with a v as the first component, replace one of the pairs (d, y), with $y \neq 0$, by (v, y).

(S3) If there are two pairs (v, y_1), (d, y_2), with y_1 greater than 2, replace them by the pairs $(v, 1)$, $(d, y_2 + y_1 - 1)$.

(S4) If there are two pairs (d, y_1), (d, y_2) with $y_1 - y_2$ being even and greater than 0, replace them by two pairs $(d, (y_1 + y_2)/2)$.

(S5) If there are two pairs (d, y_1), (d, y_2) with $y_1 - y_2$ being odd and greater than 1, replace them by the pairs $(d, (y_1 + y_2 - 1)/2)$ and $(d, (y_1 + y_2 + 1)/2)$. (Apply the ellipse trick that we used in the proof of Theorem 4.3 to prove that this and the previous instruction are really shortening rules.)

Using shortening rule S1, we easily see that the vertical length of a shortest perfect n-collection is $2(n - 1)$. Using shortening rule S2, we see that a shortest perfect n-collection has the maximal possible number of verticals. Using shortening rule S3, we convince ourselves that all pairs with first component v have vertical length 1. Using the last two shortening rules, we can show that there is a number k such that all pairs with a d as first component have vertical length k or $k + 1$. Since there are $2n$ pairs in the collection and the vertical length of the collection is $2(n - 1)$, this implies that $k = 0$ and that, consequently, a shortest perfect n-collection consists of exactly two pairs $(d, 0)$, a maximal number of pairs $(v, 1)$, and the rest pairs $(d, 1)$. Finally, it is very easy to see that the bowtie n-lacings are the only n-lacings that correspond to this shortest perfect n-collection. This completes the proof that the bowtie n-lacings are the shortest n-lacings.

We leave it as an exercise for the reader to fill in the details to make this sketch into a real proof and to modify this set-up to show that the crisscross n-lacings are the shortest dense n-lacings.

REAL LIFE ADVENTURES   **BY GARY WISE & LANCE ALDRICH**

5

Variations on the Shortest Lacing Problem

The array of eyelets in a real shoe will hardly ever be ideal. Of course, "in practice" it is a fairly straightforward exercise to churn through all lacings of a real shoe and compute the shortest lacings of this particular shoe. However, just to get a little bit of a feeling for how robust and applicable the solutions to some of our shortest shoelace problems really are, we first modify our mathematical shoe and check if and how these solutions change. It turns out that the crisscross n-lacing, as the shortest dense n-lacing, is extremely robust and, therefore, appears to be very applicable indeed. This is a (slight) generalization of a result by Misiurewicz [20]. On the other hand, we find that the bowtie n-lacings, as the solutions to the shortest shoelace problem overall, are not nearly as robust.

We also modify the way in which we lace while leaving our basic mathematical shoe unchanged. In particular, we solve the shortest shoelace problem for lacings with shoelaces that are not closed loops, for lacings consisting of several closed loops, etc.

5.1 Linear Shoes

On closer inspection of the proof of Theorem 4.1, we see that most of its arguments, such as all the shortening rules and their immediate consequences, also apply to shoes that are a lot more general than the mathematical shoes that we usually consider. Have a look at Figure 5.1. A *linear n-shoe* consists of two sets A and B of n points/eyelets such that: (1) the points of A are contained in a line a and the points of B are contained in a different line b; (2) if the two lines intersect in a point P, then all of A is contained on one side of the line a with respect to the point P or, equivalently, A is a subset of one of the two connected components of $a \setminus \{P\}$. Similarly, all of B is contained in one side of the line b with respect to the point P. The two sets A and B are the columns of eyelets. The eyelets themselves will be labeled according to one of the natural orders given by the lines a and b. In particular, if the lines intersect, then the eyelets closest to (or farthest away from) the point of intersection P will be A_1 and B_1. If the lines are parallel, choose one of the two natural "parallel" orders of the two sets; it does not matter which. Following this, all definitions of mathematical objects based on the concept of a mathematical shoe extend naturally to linear shoes. Let's call our usual setup a

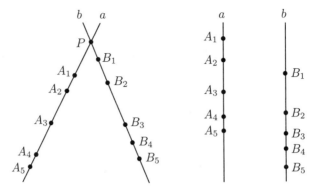

Figure 5.1. The two types of linear shoes.

classical shoe, and let's see what we can say in general about the shortest n-lacings of linear shoes.

First, using the same arguments as in the proof of Theorem 4.1, it is a straightforward exercise to prove the following theorem:

Theorem 5.1 (Shortest Lacings of Linear Shoe). *The shortest dense n-lacing of a linear shoe is the crisscross n-lacing. Every shortest n-lacing is a reduced n-lacing that does not contain three consecutive general crosses anywhere or two at the top or at the bottom (see page 31 for the definition of the term 'reduced' and page 36 for the term 'general cross').*

Figure 5.2 illustrates that a reduced n-lacing with three consecutive general crosses anywhere or two consecutive crosses at the top or bottom can be shortened. This means that no such n-lacing can be a shortest n-lacing.

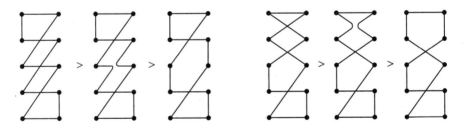

Figure 5.2. A reduced n-lacing with three consecutive general crosses anywhere or two at the top or bottom can be shortened.

We will show that for shoes that are vastly more general than linear shoes, the crisscross n-lacing is the shortest n-lacing. On the other hand, the bowtie n-lacings are not the only possible shortest n-lacings of linear shoes. Just to get an idea of some of the things that can happen, let's consider lacings of one of the simplest possible nonclassical linear shoes.

A *semi-classical shoe* is a classical shoe that has been modified by translating the right column a distance o down; see the left diagram in Figure 5.3. This means that a semi-classical shoe is determined by the vertical separation h of the eyelets and this offset o (remember that the horizontal separation of the two columns is always 1).

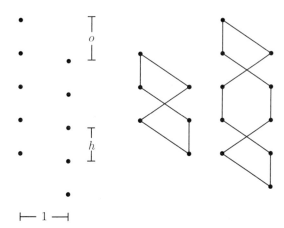

Figure 5.3. A semi-classical shoe and the shortest 3- and 5-lacings of semi-classical shoes with $h = o$.

In a semi-classical shoe with an odd number of eyelets and $h = o$, it is easily seen that the shortest n-lacing is not a bowtie n-lacing but rather a *skew-bowtie n-lacing*; the middle and right diagrams in Figure 5.3 show the skew-bowtie 3- and 5-lacings. It should be clear what the skew-bowtie n-lacing looks like for general odd n. The argument for concluding that the skew-bowtie n-lacing is the shortest n-lacing for the given choices of n and $h = o$ is the same as the one we used to conclude that the bowtie n-lacings are the shortest n-lacings—the skew-bowtie n-lacing contains the maximum possible number of 1-verticals and all its general crosses are crosses. We only note that, in general, for n odd some numerical experiments suggest that, for fixed h, the shortest n-lacings of a semi-classical shoe are the bowtie n-lacings for small o and that from a certain value of o onward the shortest n-lacing is the skew-bowtie n-lacing.

For even n, numerical experiments suggest that the shortest n-lacing depends on both h and o in a more complicated manner. The case $n = 4$ is symptomatic. Here, it turns out that for small h (up to about $h = 0.95$) the shortest n-lacing is the bowtie 4-lacing, no matter what we choose o to be. However, for larger h, things change. In this case, the shortest n-lacing is also the bowtie 4-lacing for small o. However, if o is in an interval that starts a little bit before h and ends a little bit before $3h$, then the smallest 4-lacing is the one in the second diagram in Figure 5.4. The lacings in the diagram are drawn to scale, and here $o = 2h = 2$. Since the bowtie 4-lacing and this 4-lacing differ in only the configuration of segments drawn thick, we can use several copies of the configuration in the second 4-lacing to build n-lacings for all even $n > 2$ that are shorter than the bowtie n-lacings, as demonstrated in the diagrams on the right for the case $n = 6$. The resulting n-lacings are never the shortest n-lacings since they contain lots of consecutive general crosses, which can be shortened; see Figure 5.4. For larger o the smallest 4-lacing is, again, the bowtie 4-lacing.

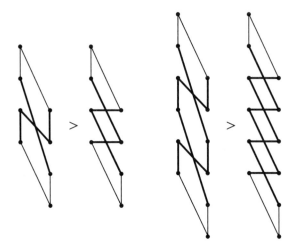

Figure 5.4. The shortest 4-lacing of a semiclassical shoe with $2h = o = 2$ and, derived from it, a 6-lacing of a semiclassical shoe with $2h = o = 2$ that is shorter than the bowtie 6-lacing of this shoe.

5.2 Framed Shoes

As an application of Theorem 5.1, let's consider the following scenario: Start with a natural number $n > 2$ and a rectangle whose sides are horizontal and vertical. Call its four corners A_1, B_1, A_n, B_n, as in Figure 5.5. Now, turn this rectangle into a linear shoe by punching $n - 2$ more distinct eyelets on the left side and $n - 2$ distinct eyelets on the right side of the rectangle, and lace this shoe any way you want. Let's call the lacings that we arrive at in this way *framed n-lacings*. Is there a shortest framed n-lacing?

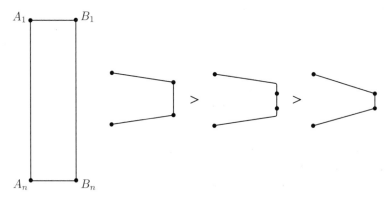

Figure 5.5. The rectangular frame for a framed shoe on the left and a shortening rule for framed n-lacings on the right.

Lemma 5.2 (Existence). *For $n > 2$ there is no shortest framed n-lacing.*

Proof. Let's assume that there is a shortest framed n-lacing. As a consequence of Theorem 5.1, it contains the top and bottom horizontals. Hence, it cannot contain both the verticals A_1A_n and B_1B_n. If it contains one of these long verticals, say A_1A_n, then it also has to contain a shorter vertical in column B. However, it cannot contain any short verticals, because by moving one of the ends of such a vertical towards the other, you can always shorten a framed n-lacing, as illustrated in the right diagram of Figure 5.5. This means that our shortest framed n-lacing is a dense n-lacing. Hence, by Theorem 5.1, our shortest framed n-lacing is a crisscross n-lacing. However, it is clear that a crisscross n-lacing can always be shortened by replacing any one of its crosses by two verticals. This is a contradiction to our assumption, which means that there is no shortest framed n-lacing. $\square$

On the other hand, we can prove the following very satisfying theorem using *Heron's Principle*, one of mathematics' all-time classics; see, for example, [15].

Theorem 5.3 (Shortest Dense Lacings of Framed Shoe). *The shortest dense framed n-lacing is the crisscross n-lacing of the classical shoe that we arrive at by punching in the extra eyelets equally spaced.*

Proof. By Theorem 5.1, we know that for any possible choice of punching the eyelets, the crisscross n-lacing is the shortest dense n-lacing. Therefore, it suffices to show that the shortest framed crisscross n-lacing is that of the classical shoe. We really need to check two things here. First, that a shortest framed crisscross n-lacing exists and, second, that this shortest crisscross n-lacing is that of the classical shoe.

As a consequence of Theorem 5.1, all lacings under consideration share the top and bottom horizontals. This means that, when we compare the lengths of two such lacings, we can restrict ourselves to comparing the lengths of the lacings minus the two horizontals. So, let's focus on an arbitrary framed crisscross n-lacing and delete the bottom and top horizontals; see Figure 5.6. This leaves us with two connected parts. Grab the part that contains the top left eyelet. If n is odd, reflect it in the vertical mirror axis of our frame. If n is even, as in the diagrams, don't do anything at this point. Now, shift what you are holding on to down until its top end connects to the bottom end of the other part; see the fourth diagram in Figure 5.6. We also encase the resulting connected zigzag path in a rectangular frame double the vertical size of the one in which our framed n-lacings live. Now, it is easy to see that the zigzag path has the following properties:

1. It consists of $2(n-1)$ straight line segments whose endpoints are contained in the two vertical sides of our long frame.
2. The endpoints of a segment are contained in different sides.
3. The two endpoints of the path are the top and bottom right corners of the frame.

Let's call a continuous zigzag path in the frame that satisfies conditions 1-3 above an n-*lightning*. Note that in an n-lightning some of the endpoints of segments may coincide. We want to show that there exists a shortest n-lighting and that this shortest n-lighting is the one that arises from the crisscross n-lacing of a classical shoe, as described above. Of course, this will then imply immediately that this particular crisscross n-lacing is the shortest framed one.

We start by assuming that there exists a shortest n-lightning. We want to apply Heron's principle, which says the following: Given two points A and B on one side of a line l, the shortest path from A to B via l is the one that a light ray

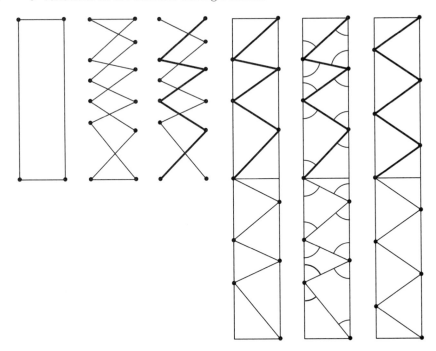

Figure 5.6. Shortening a framed crisscross n-lacing.

emanating from A and reflected off l would take. This implies, in particular, that the two segments that this path consists of make equal angles with l. This in turn immediately implies that in a shortest n-lightning all the angles marked in the fifth diagram of Figure 5.6 are equal, that is, that the shortest n-lightning is the n-lightning that arises from a classical shoe shown in the sixth diagram.

We finish the proof by verifying the existence of an n-lightning of minimal length. If k is the length of our frame, we can coordinatize every endpoint by its distance from the top corner of the frame above it. Furthermore, it is easy to check that every n-lightning has $n-1$ endpoints on the left side and n endpoints on the right side (two of which are always the top and bottom corners). This means that if we note the coordinates of the endpoints different from the ends of the lightning, as we traverse a lightning from top to bottom, we arrive at a sequence of $2n-3$ numbers in the interval $[0, k]$. On the other hand, every such sequence corresponds in a unique way to an n-lightning. This means that we can identify the set of n-lightnings with the compact set $[0, k]^{2n-3}$. Let f be the function $[0, k]^{2n-3} \to \mathbf{R}$ that assigns to every n-lightning its length. Then it is clear that f is a continuous real-valued function on a compact set. As such, f has a minimum. This means that there really exists an n-lightning of minimal length. $\qquad\square$

5.3 General Shoes

Misiurewicz proved in [20] that the crisscross n-lacing is the shortest among all the dense n-lacings containing the top horizontal of a very general kind of shoe. In the following, we give an alternative proof for this result and show that the extra condition involving the horizonal segment is superfluous.

Let $A_1, A_2, \ldots, A_n, B_1, B_2, \ldots, B_n$ be $2n$ distinct points/eyelets in the plane such that the *cross-straight* shortening rule depicted in Figure 5.7 is valid for all possible choices of $i < j$ and $k < l$, that is,

$$|A_i B_l| + |A_j B_k| > |A_i B_k| + |A_j B_l|.$$

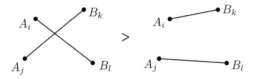

Figure 5.7. The cross-straight shortening rule for a general n-shoe.

In particular, this is the case if, as in the diagram, the four eyelets A_i, A_j, B_k, and B_l form the vertices of a nondegenerate quadrangle such that the segments $A_i B_k$ and $A_j B_l$ on the right side of this inequality are opposite sides of this quadrangle, and the segments $A_i B_l$ and $A_j B_k$ on the left side are its diagonals. Clearly, this is the case in our usual set-up, in the case of linear shoes, as well as in a vast variety of other arrangements of the $2n$ eyelets.

We call an arrangement of $2n$ eyelets that satisfies the condition above a *general n-shoe*. For the rest of this section, we will assume that n-lacings and all associated terms are defined with respect to general n-shoes. For example, Figure 5.8 shows the crisscross n-lacing of a general n-shoe.

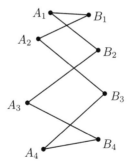

Figure 5.8. The crisscross 4-lacing of a general 4-shoe.

Theorem 5.4 (Shortest Dense Lacings of General Shoes). *The crisscross n-lacing of a general n-shoe is the shortest dense n-lacing of this shoe.*

Proof. To prove this result in our usual setting, we used the set of shortening rules in Figure 4.6. It turns out that these shortening rules extend to our more general setting. To show this, we deduce every one of them by applying the cross-straight shortening rule one or two times, as summarized in Figure 5.9.

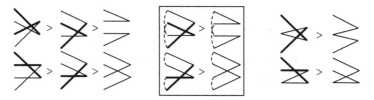

Figure 5.9. Deducing the shortening rules in Figure 4.6 by applying the cross-straight shortening rule to pairs of thick segment.

Note that at every step we indicate which cross of segments gets replaced by drawing the two segments that form the cross thick. Also, it should be clear that to interpret one of the pairs of zigzags in our more general setting, the feet are supposed to be labeled by eyelets in one column whose indices increase from top to bottom. Similarly, the horns are supposed to be labeled by eyelets in the other column whose indices also increase from top to bottom.

Now that all these shortening rules are also available in the more general setting, our proof of the second part of Theorem 4.1 can be easily generalized to the more general setting that we are concerned with at the moment.

Since there is only one dense 2-lacing, there is nothing to prove in the case that $n = 2$. Therefore, let $n > 2$, and let S be a shortest dense n-lacing. We start by proving that the top horizontal $A_1 B_1$ is contained in S by considering the possible configurations of pairs of zigzags in such a dense n-lacing that have A_1 as one of their feet and B_1 as one of the horns. We do this by labeling the top-most foot and the top-most horn of all possible reduced configurations of pairs of zigzags in S with A_1 and B_2, respectively; see Figure 5.10

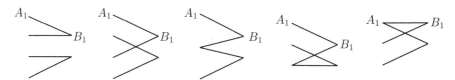

Figure 5.10. Labelling the top-most foot and horn in all possible reduced configurations of zigzags with A_1 and B_1 shows that the horizontal $A_1 B_1$ is contained in any shortest dense n-lacing.

As you can see, the top horizontal is always present, which implies that this horizontal is part of S. To see that the diagonal $A_1 B_2$ is contained in S, we consider a pair of zigzags in S that contains the foot A_1 and B_2, and does not contain B_1. Such a configuration always exists and has A_1 as its top-most foot and B_2 as its top-most horn. We conclude as before that $A_1 B_2$ is contained in this configuration and, therefore, also in S. Similarly, we see that $B_1 A_2$ is contained in S. Of course, arguing in the same way as to this point, we also see that the bottom horizontal and the diagonals $A_n B_{n-1}$ and $B_n A_{n-1}$ are contained in S. If $n = 3$, this completes the proof. Otherwise, we continue by showing that the diagonal $A_2 B_3$ is contained in S by considering all possible pairs of zigzags that contain the foot A_2 and B_3 but none of the eyelets A_1, B_1, and B_2. From this point on things repeat, and it easily follows that S is the crisscross n-lacing.   □

Note that the application of the cross-straight shortening rule itself to a lacing does not necessarily result in another lacing. What can go wrong is that we may end up with more than one loop. This means that if we want to use the cross-straight shortening rule to prove the theorem above, it is not always possible to apply it directly. Instead, we have to apply it indirectly, for example, as we have done in the previous proof, via our set of shortening rules in Figure 5.9 that are based on this shortening rule.

5.4 Open Lacings

For the rest of this chapter, we will switch back to lacing classical shoes. An *open n-lacing* is an n-lacing minus one of those among its segments that is preceded and followed by diagonals. This means that an open n-lacing has two end eyelets, that the segments ending in these distinguished eyelets are diagonals, and that, just as in the case of n-lacings, every eyelet contributes at least once towards pulling the two sides of the shoe together. A *dense* open n-lacing is an open n-lacing that arises from a dense n-lacing. Clearly, every dense n-lacing corresponds to exactly $2n$ dense open n-lacings. Figure 5.11 shows two examples of open 5-lacings that generalize in a straightforward way to open n-lacings for all n.

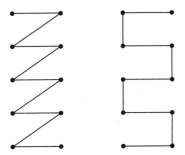

Figure 5.11. Examples of open 5-lacings—an open zigzag and an open zigsag 5-lacing.

We call the open n-lacings that correspond to the left-hand diagram and its vertical mirror image the *open zigzag n-lacings*.[1] We call the open n-lacings that correspond to the right-hand diagram and its vertical mirror image the *open zigsag n-lacings*.

Theorem 5.5 (Shortest Open Lacings of Classical Shoes). *The two open zigzag n-lacings are the shortest dense open n-lacings. The two open zigsag n-lacings are the shortest open n-lacings.*

Note that these two open solutions to the shortest shoelace problem are both 'straight' and 'simple'.

---

[1] Archeological finds show that in the past open zigzag n-lacings were used to lace real shoes. Furthermore, it is worth noting that a seam that is produced by the simplest way of stitching two parts of a shoe together looks very much like an open zigzag lacing. This suggests considering lacings as open seams.

Proof. The shortest segments that can be used for building a dense open n-lacing are the n horizontals. The next shortest segments are the 1-diagonals. Clearly, the open zigzag n-lacings are the only dense open n-lacings that contain all n horizontals and the rest 1-diagonals. Hence, they are the shortest dense n-lacings.

The shortest segments that can be used for building open n-lacings are the horizontals and the 1-verticals. The open zigsag n-lacings each contain n horizontals and $n-1$ verticals. We first show that these are also the maximal numbers of horizontals and verticals in any open n-lacing. In terms of horizontals this is clear. Two 1-verticals in an open n-lacing cannot share an eyelet. Therefore, we conclude, as in the case of n-lacings, that the maximal number of 1-verticals contained in one of the columns is $(n-1)/2$ in the odd case and $n/2$ in the even case. Hence, in the odd case, $n-1$ is indeed the maximal possible number of 1-verticals. In the even case, let's assume that there is an open n-lacing that contains the maximal possible number of verticals in both columns, that is, n verticals in total. Then in the n-lacing that completes the open lacing, verticals and diagonals alternate. However, if we remove a diagonal from this n-lacing, then the two ends of what is left are connected to the rest by verticals and we do not get an open n-lacing in this way. This means that since, as in the odd case, there are open n-lacings with $n-1$ verticals, the maximal number of 1-verticals in an open n-lacing is $n-1$.

We conclude that the zigsag n-lacings are the shortest open n-lacings because the zigsag n-lacings are the only open n-lacings that contain $n-1$ verticals and all n horizontals.   □

5.5 Multi-Loop Lacings

An n-lacing consists of just one loop. What if we allow lacing using several loops, but leave the rest of our requirements unchanged? Let's call lacings like this *multi-loop n-lacings*. We call a loop consisting of two copies of the same 1-horizontal a *0-loop* and a loop consisting of two 1-verticals and two horizontals a 1-*loop*; see Figure 5.12.

Figure 5.12. A 0-loop and a 1-loop.

It is possible, using the same kind of simple arguments as in the previous section, to prove the following result:

Theorem 5.6 (Shortest Multi-Loop Lacings of Classical Shoes). *If the stretch of the shoe h is greater than 1, then the shortest multi-loop n-lacing consists of n 0-loops; see the left diagram in Figure 5.13. If $h < 1$, then a multi-loop n-lacing is a shortest one if it consists of n/2 1-loops if n is even and $(n-1)/2$ 1-loops and one 0-loop if n is odd; see the right diagram in Figure 5.13. If $h = 1$, then a multi-loop n-lacing is a shortest one if it consists only of 0-loops and 1-loops.*

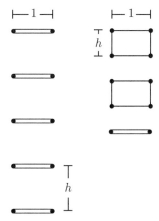

Figure 5.13. Shortest multi-loop 5-lacings for long and short shoes.

The last straw is very often a shoelace.

Figure 5.14. Some examples of decorative lacings.

6
The Longest Lacings

In the last two chapters, we found that the solutions of the shortest shoelace problem in all kinds of different natural settings are basically unique and of a very simple structure. It is natural to ask whether the same is true for the solutions of the longest shoelace problem. To start with, my aim was not to come up with a set of solutions that is as comprehensive as that for the shortest shoelace problem. However, soon things got out of hand, and I ended up deriving the most appealing among these solutions and complementing these by computer-experiment-based conjectures about the other solutions. My main results are summarized in Figure 6.1. In this Venn diagram the conjectured solutions are marked with a question mark.

6.1 Summary of Results and Conjectures

One of the most interesting results of this chapter is the characterization of the zigzag n-lacings as the longest simple n-lacings. This means that in the class of n-lacings that are both dense and simple the classical lacings are the extremes in terms of length—just remember that, by Theorem 4.1, the crisscross n-lacing is the shortest dense-and-simple n-lacing.

We introduce the so-called devil n-lacing and show that it is the longest dense n-lacing. The devil n-lacing, as the solution of the longest dense shoelace problem, turns out to be just as robust as the crisscross n-lacing, the solution of the shortest dense shoelace problem.

The shortest n-lacings in the different classes of n-lacings considered by us do not change when we modify the stretch of our classical shoe and are basically unique. On the other hand, we show that the longest n-lacing does depend on the stretch of this shoe and can be one of at least two very different types of n-lacings. In particular, we prove that the devil n-lacing is the longest n-lacing for short shoes, and we conjecture that the so-called angel n-lacings are the longest n-lacings for long shoes.

In the next section, we prove that the longest superstraight n-lacings are exactly those superstraight n-lacings that contract to the longest one-column n-lacings. In the same section, we also prove that the longest straight n-lacings are the so-called straight devil n-lacings.

In a final section, we characterize the n-lacings of longest vertical length.

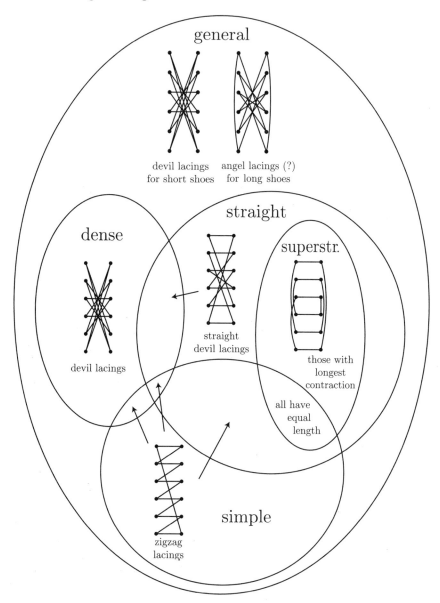

Figure 6.1. The longest lacings in the different classes of n-lacings (? = conjecture).

6.2 The Longest Simple n-Lacings

The following theorem is the first neat characterization of the zigzag n-lacings that we come across in this book. We will derive a second one in the following chapter.

Theorem 6.1 (Longest Simple). *The longest simple n-lacings are the zigzag n-lacings.*

Proof. This proof is very similar to our proof that the star n-lacings are the shortest n-lacings that are both dense and straight; see page 39.

It is clear that this theorem is true in the case $n = 2$. So, let $n > 2$. The collection V of the vertical lengths of the segments in a simple n-lacing has the following properties:

(V1) V consists of $2n$ nonnegative integers.
(V2) V contains at most n 0s.
(V3) The sum of the integers in V is $2(n-1)$.

We call a collection V of nonnegative integers that has properties V1, V2, and V3 a *simple exploded n-lacing*. Note that not every simple exploded n-lacing necessarily arises from a simple n-lacing. Moreover, a simple exploded n-lacing may arise from a number of very different simple n-lacings.

We define the length of a simple exploded n-lacing V to be the sum

$$\sum_{s \in V} \sqrt{1 + (hs)^2},$$

where h is the stretch of the underlying mathematical shoe. Clearly, the length of a simple exploded n-lacing that arises from a simple n-lacing is greater than or equal to the length of the simple n-lacing. The two lengths are equal if and only if the n-lacing is also dense.

The *simple exploded zigzag n-lacing* is the simple exploded n-lacing that arises from the zigzag n-lacings. It consists of n 0s, $n-1$ 1s and one $n-1$. We now show that the simple exploded zigzag n-lacing is the longest simple exploded n-lacing. Let S be a longest simple exploded n-lacing. If S does not contain n 0s, then we replace two of its elements e and f that are not equal to 0 by a 0 and an $e + f$ to arrive at a new simple exploded n-lacing. Using the same 'ellipse-trick' as in the proof of Theorem 4.3, we can convince ourselves that this new simple exploded n-lacing is longer than the one we started with, which is a contradiction. We conclude that our longest simple exploded n-lacing S contains n 0s. Similarly, if S does contain less than $n-1$ 1s, then there are at least two elements e and f of S that are greater than 1. Replacing these elements by a 1 and an $e + f - 1$ gives a longer simple exploded n-lacing (we once more use the ellipse trick). This is again a contradiction, which implies that S also contains at least $n-1$ 1s. Because of V3 this also implies that the one element among the $2n$ elements of S that we have not accounted for yet is equal to $n-1$. This means that the longest simple exploded n-lacing is indeed the simple exploded zigzag n-lacing.

Remember that the length of a simple exploded n-lacing that arises from a simple n-lacing is greater than or equal to the length of the n-lacing and that the two lengths are equal if and only if the n-lacing is dense. Now, it is easy to see that the zigzag n-lacings are the only dense-and-simple n-lacings that give rise to the simple exploded zigzag n-lacing. We conclude that the zigzag n-lacings are the longest simple n-lacings.   □

6.3 The Longest Dense and General n-Lacings

In Chapter 5, we succeeded in identifying the crisscross n-lacing of a general n-shoe as the shortest dense n-lacing of this shoe. We did this using the set of shortening rules in Figure 5.9, which was derived from the cross-straight shortening rule in

Figure 5.7. In the following, we want to move in the opposite direction and, using a set of "lengthening rules", identify the so-called devil n-lacing as the longest dense n-lacing of a given general n-shoe. We recommend that you review Section 5.3 on lacing general shoes before you read on, as we will assume familiarity with the terminology and arguments introduced there.

The "inverse" of the cross-straight shortening rule for general n-shoes, considered in the last chapter, is the *straight-cross lengthening rule* summarized in Figure 6.2.

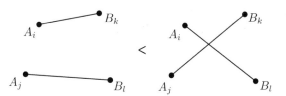

Figure 6.2. The straight-cross lengthening rule for general shoes.

Let $n > 2$. Using the straight-cross lengthening rule, we can derive the set of six lengthening rules in Figure 6.3 that contains all possible configurations of pairs of zigzags of the first kind. Of course, when we say "all", we mean all, up to forming horizontal and vertical mirror images. Note that we arrive at these lengthening rules by reshuffling the diagrams in Figure 5.9.

Applied to a lacing, the straight-cross lengthening rule may or may not result in a lacing. What can go wrong is that we may end up with more than one loop. However, whenever we apply one of the lengthening rules in Figure 6.3 to an n-lacing of a general n-shoe, we arrive at a longer n-lacing of this shoe.

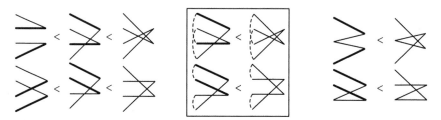

Figure 6.3. Applying the straight-cross lengthening rule to the thick pairs of segments in the diagrams yields six lengthening rules.

We conclude that a longest dense n-lacing of a general n-shoe can only contain the configurations of pairs of zigzags in Figure 6.4.

We'll now deduce the longest dense n-lacings for general n-shoes using the straight-cross lengthening rule and the six lengthening rules in Figure 6.3. Just as with these lengthening rules, all our arguments will apply to all general n-shoes. Therefore, we can restrict ourselves to deducing the longest n-lacing for one fixed n-shoe because, described in terms of A_is and B_is, this longest n-lacing will be the same for all general n-shoes. *In the following, we will therefore restrict ourselves to working in our usual setting; that is, choose the fixed n-shoe to be a classical shoe.*

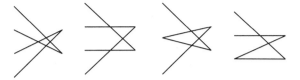

Figure 6.4. The possible configurations of pairs of zigzags in a longest dense n-lacing, up to reflections in vertical and horizontal axes.

The *devil n-lacing* consists of the two main diagonals A_1B_n, B_1A_n and,

- for n even, all possible segments passing through the centers of the middle two horizontals $A_{n/2}B_{n/2}$ and $A_{n/2+1}B_{n/2+1}$ (see the three diagrams on the left in Figure 6.5);
- for n odd, all possible segments passing through the center of the rectangle

$$A_{(n+1)/2}A_{(n-1)/2}B_{(n-1)/2}B_{(n+1)/2},$$

and those passing through the center of the rectangle

$$A_{(n+1)/2}A_{(n+3)/2}B_{(n+3)/2}B_{(n+1)/2}$$

(see the two diagrams on the right in Figure 6.5).

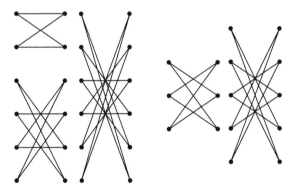

Figure 6.5. The devil n-lacings for small n.

As you can see, a devil n-lacing is constructed from an empty shoe step by step using a greedy algorithm such that at any step you always add the longest possible diagonal.[1]

Theorem 6.2 (Longest Dense). *The devil n-lacing (of a general n-shoe) is the longest dense n-lacing (of this general n-shoe).*

Proof. Since the devil 2-lacing is the only dense 2-lacing, there is nothing to prove in the case $n = 2$. For $n = 3$, there are four dense 3-lacings (up to symmetries of the underlying mathematical shoe); see Figure 3.3. Using only the straight-cross lengthening rule, we can show that the devil 3-lacing is the longest dense 3-lacing; see Figure 6.6.

---

[1] Here "possible diagonal" is also supposed to mean that after inserting a diagonal it is actually still possible to complete what you have constructed so far to a lacing.

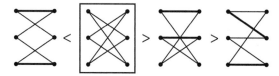

Figure 6.6. Applying the straight-cross lengthening rule to a pair of thick segments in one of the 3-lacings yields a longer 3-lacing.

From now on, let $n > 3$. Let l be a longest dense n-lacing. We first want to convince ourselves that the main diagonals A_1B_n and B_1A_n are contained in l; see the left-hand diagram in Figure 6.7. We can restrict ourselves to showing this for A_1B_n. We consider the possible configurations of pairs of zigzags such that A_1 is one of the feet of this configuration and B_n is one of the horns. We do this by labeling the top-most foot of each of the configurations in Figure 6.3 (and of the images of these configurations under horizontal mirror reflections) with A_1 and the bottom-most horn with B_n. We find that the main diagonal A_1B_n is always present, no matter which way we label. A similar argument shows that the second main diagonal B_1A_n is also contained in l.

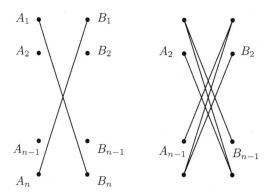

Figure 6.7. Building a devil n-lacing.

We want to show that the four second longest diagonals A_1B_{n-1}, B_1A_{n-1}, B_2A_n, and A_2B_n are all part of a longest dense n-lacing; see the right-hand diagram in Figure 6.7.

We can restrict ourselves to showing this for A_1B_{n-1}. There is a configuration of pairs of zigzags in l such that A_1 is one of the feet, B_{n-1} is one of the horns, and the long diagonal A_1B_n is not contained in the configuration. This means that B_n is not a horn of this configuration. Therefore, it is clear that A_1 is the top-most foot and B_{n-1} the bottom-most horn of the configuration, and we conclude, as before, by labeling all possible configurations in Figure 6.3 that A_1B_{n-1} is contained in our configuration.

Next, as you can see in the second diagram in Figure 6.7, neither the diagonal A_2B_{n-1} nor the diagonal B_2A_{n-1} can be contained in our lacing, because inserting one of these segments would create more than one loop.

Therefore, if $n = 4$, the only way to complete our partial lacing to a dense 4-lacing is to insert the horizontals $A_2 B_2$ and $A_{n-1} B_{n-1} = A_3 B_3$. The resulting 4-lacing is the devil 4-lacing. This shows that the longest dense 4-lacing is the devil 4-lacing.

Let $n = 5$. If the horizontal $A_2 B_2$ was contained in l, then the horizontal $A_{n-1} B_{n-1}$ could not be contained in l because it would close the lacing prematurely. Therefore, l would be the 5-lacing shown in the middle diagram of Figure 6.8, which is impossible since this lacing can be lengthened by applying the straight-cross lengthening rule to the thick segments. It follows that the devil 5-lacing on the right is the longest dense 5-lacing.

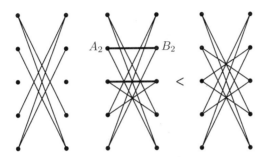

Figure 6.8. Building a devil 5-lacing.

Let $n > 5$. Then we can argue as before that the four next longest (possible) diagonals $A_2 B_{n-2}$, $B_2 A_{n-2}$, $A_3 B_{n-1}$, and $B_3 A_{n-1}$ are contained in l. Following this, we argue as in the cases $n = 4$ and $n = 5$ that both in the cases $n = 6$ and $n = 7$ the devil lacings are the longest dense n-lacings.

From there on things repeat in the obvious manner, which allows us to conclude that for any n the longest dense n-lacing is the devil n-lacing.   □

The rest of this section concerns only our usual setting; that is, we will be only dealing with classical n-shoes.

Assume that h, the stretch of the shoe, is chosen so small that every possible diagonal segment is longer than any possible vertical segment. Then we can show that the longest n-lacings are dense.

To see this, remember that if an n-lacing has a vertical whose endpoints are eyelets in column A, then it also contains a vertical whose endpoints are contained in column B; see Lemma 1.1. These two verticals may be tied into the overall lacing in essentially two different ways, as indicated by the dashed curves in the left two diagrams in Figure 6.9. In both cases, we can replace the two verticals by two diagonals, as indicated in the right two diagrams. This results in longer lacings. In the first replacement, we apply a quarter-turned version of the straight-cross lengthening rule. In the second replacement, the lacing on the right is longer than the lacing on the left because, by assumption, any diagonal is longer than any of the verticals. We conclude that a longest n-lacing is dense. This yields the following corollary of Theorem 6.2:

Corollary 6.3 (Longest for Short Shoes). *For short shoes, the devil n-lacing is the longest n-lacing overall.*

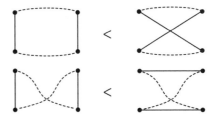

Figure 6.9. For small h the longest n-lacings are dense.

We did not succeed in figuring out what exactly the longest n-lacings are for the different possible choices of n and h. However, we can show that, for large h, the devil n-lacings are not among the longest n-lacings. For example, Figure 6.10 shows the longest n-lacings for some small values of n and large h, up to reflections in horizontal and vertical axes (this is based on computer experiments).

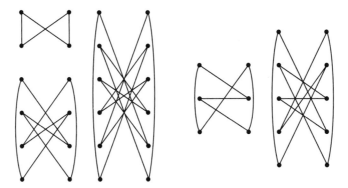

Figure 6.10. The angel n-lacings for small n.

Let's define the *angel n-lacings* as the n-lacings that are the natural generalizations of the lacings in Figure 6.10. Note that there is exactly one angel n-lacing for n even and two for n odd (one the mirror image of the other). Furthermore, for $n > 2$, an angel n-lacing and a devil n-lacing differ only in four segments; see Figure 6.11.

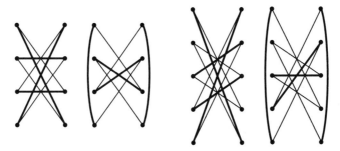

Figure 6.11. An angel n-lacing and a devil n-lacing differ only in four segments.

Theorem 6.4 (Relative Lengths of Devils and Angels). *Let $n \geq 2$, and let $A(n, h)$, $D(n, h)$ be the lengths of the angel and devil n-lacings, respectively. Then, for fixed n, there is exactly one $h_n > 0$ such that $A(n, h_n) = D(n, h_n)$. For $h < h_n$, $D(n, h) > A(n, h)$, whereas for $h > h_n$, $A(n, h) > D(n, h)$.*

Proof. For $n = 2$ this theorem is clearly true (and $h_2 = 1$). Let $n > 2$, and define the function

$$c_n : \mathbf{R}^{0+} \to \mathbf{R} : h \mapsto D(n, h) - A(n, h).$$

We want to show that c_n has exactly one positive zero h_n and that c_n is positive for $0 < h < h_n$ and negative for $h > h_n$.

First, let n be even. Then

$$c_n(h) = 2\left(\sqrt{1 + (n-1)^2 h^2} + 1 - (n-1)h - \sqrt{1 + h^2}\right).$$

Clearly, $c_n(0) = 2$ and $\lim_{h \to \infty} c_n(h) = -\infty$. Therefore, the equation $c_n(h) = 0$ has at least one positive solution. For $h > 0$,

$$c_n'(h) = 2\left(\frac{h(n-1)^2}{\sqrt{1 + (n-1)^2 h^2}} - (n-1) - \frac{h}{\sqrt{1 + h^2}}\right)$$

$$= 2(n-1)\left(\frac{1}{\sqrt{\frac{1}{(n-1)^2 h^2} + 1}} - 1\right) - \frac{2}{\sqrt{\frac{1}{h^2} + 1}}.$$

Since the first and second terms in the last expression are clearly both negative, we conclude that $c_n'(h) < 0$. Consequently, the equation $c_n(h) = 0$ has a unique solution, h_n.

Now, let n be odd. Then

$$c_n(h) = 2\sqrt{1 + (n-1)^2 h^2} + 2\sqrt{1 + h^2} - \sqrt{1 + 4h^2} - 1 - 2(n-1)h.$$

Clearly, $c_n(0) = 2$ and $\lim_{h \to \infty} c_n(h) = -\infty$. We will show that for positive h the derivative of this function is negative. This implies that the function is strictly decreasing and, consequently, that this function has exactly one positive zero. The derivative of c_n is

$$\frac{2h}{\sqrt{1 + h^2}} + \frac{-4h}{\sqrt{1 + 4h^2}} - 2(n-1) + \frac{2h(n-1)^2}{\sqrt{1 + h^2(n-1)^2}}$$

$$= 2\left(\frac{1}{\sqrt{\frac{1}{h^2} + 1}} - \frac{1}{\sqrt{\frac{1}{4h^2} + 1}}\right) + 2(n-1)\left(\frac{1}{\sqrt{\frac{1}{(n-1)^2 h^2} + 1}} - 1\right).$$

Since the expressions in the brackets are both clearly negative for positive h, we conclude that the derivative of c_n is negative.   □

Based on the last theorem and our numerical experiments, we venture the following conjecture:

Conjecture 6.5 (Longest). Depending on h, the longest n-lacings are as follows:

- The devil n-lacing for $h < h_n$.
- The devil n-lacing and the angel n-lacing(s) for $h = h_n$.
- The angel n-lacings for $h > h_n$.

Here h_n is defined as in the last theorem.

6.4 The Longest Straight and Superstraight n-Lacings

Remember that, apart from the horizontals, superstraight n-lacings contain only verticals. This means that the length of a superstraight n-lacing is equal to the length of its contraction plus n. We gave explicit constructions for the longest one-column n-lacings in Chapter 2 and demonstrated that, for even $n > 2$, there are exactly two superstraight n-lacings that contract to any one of these longest one-column n-lacings, one being the vertical mirror image of the second. Furthermore, the bowtie 2-lacing is the only superstraight 2-lacing. This immediately implies the following result:

Theorem 6.6 (Longest Superstraight). *The longest superstraight n-lacings are exactly those superstraight n-lacings that contract to the longest one-column n-lacings.*

Since all simple-and-superstraight n-lacings clearly have the same length, it does not really make much sense to speak of the longest such n-lacings.

A *straight devil n-lacing* is built up segment by segment from a shoe that has been fitted with all horizontals such that at any stage we fit in the longest possible segment; see Figure 6.12. There are two straight devil n-lacings for every $n > 2$, one the vertical mirror image of the other. So, just as the devil n-lacings, these lacings are constructed using a greedy algorithm.

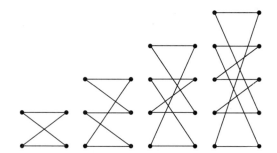

Figure 6.12. The straight devil 2-, 3-, 4-, and 5-lacings.

Theorem 6.7 (Longest Straight). *The straight devil n-lacings are the longest straight n-lacings.*

Proof. It is clear that this theorem is true for $n = 2$. For $n = 3$, all straight n-lacings are simple; see Figure 3.3. Therefore, by Theorem 6.1, the longest straight 3-lacings are the zigzag 3-lacings, which are the straight devil 3-lacings. We conclude that the theorem is true in the case $n = 3$. In the following, we will assume that $n > 3$.

Let l be a straight n-lacing. If l is not dense, then it is clear that its diagonalization (see Section 2.4) is a straight n-lacing that is longer than l. We conclude that a longest straight n-lacing is dense.

Next, we want to prove that the contraction of a longest straight n-lacing is one of the longest one-column n-lacings. To understand the following arguments you have to review Section 2.3, in which we construct the longest one-column n-lacings. Let l be dense-and-straight, and let's assume that its contraction C is not one of the longest one-column n-lacings. Then the contraction C contains two nonoverlapping segments s and s' and can be lengthened by replacing these two segments by two segments w and w' using one of the two lengthening rules illustrated by the top two diagrams in Figure 6.13 (this part of the diagram is just Figure 2.5).

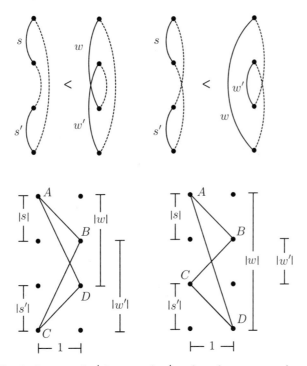

Figure 6.13. Replacing s and s' by w and w' to lengthen a one-column n-lacing, and comparing the lengths of the corresponding segments in the n-lacing setting.

Let's call the resulting one-column n-lacing C', pick one of the dense-and-straight n-lacings that contracts to C', and call it l'. We want to show that l' is longer than l, which then immediately implies that a longest straight n-lacing contracts to a longest one-column n-lacing. From the way l and l' are built, it is clear that what we have to prove is that

$$|AB| + |CD| < |AD| + |CB|,$$

where the points A, B, C, and D are as in one of the two bottom diagrams in Figure 6.13. Here the left or right diagram applies depending on which of the two lengthening rules we used before. In both cases, it is easy to prove that this inequality is true using the straight-cross lengthening rule that we introduced in the previous section; see Figure 6.2.

Finally, we are ready to prove that the straight devil n-lacings are the longest straight n-lacings. We first do this for n even. Have a look at Figure 6.14. On the left, you see a straight devil 6-lacing. The second diagram shows the contraction of this 6-lacing. As you can see, it is indeed one of the longest one-column 6-lacings. To construct the third diagram from the one-column lacing, just grab the top part of the one-column lacing and move it one unit to the right while keeping all segments straight. The third diagram is the devil 3-lacing of a linear shoe that shares the bottom three eyelets on the left and the top three eyelets on the right with our original shoe. Furthermore, the segments in the 3-lacing are just the nonhorizontal segments in the straight devil n-lacing (rearranged). Therefore, the length of this 3-lacing is exactly the length of the straight devil lacing that we started with minus the total length of the horizontals.

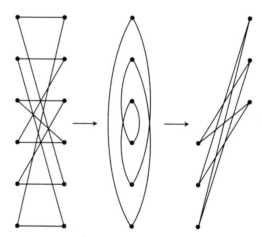

Figure 6.14. Turning a dense-and-straight n-lacing, n even, that contracts to a longest one-column n-lacing into a dense $(n/2)$-lacing (of a different linear shoe).

In general, we construct a new (linear) shoe whose eyelets are the bottom $n/2$ eyelets in the left column of our original shoe and the top $n/2$ eyelets of the right column of this shoe. Then the construction we just introduced establishes a two-to-one correspondence between the dense-and-straight n-lacings that contract to one of the longest one-column n-lacings and the dense $(n/2)$-lacings of the new shoe. This correspondence is two-to-one because both one of these special dense-and-straight n-lacings and its vertical mirror image correspond to the same dense $(n/2)$-lacing of the new shoe. The length of one of the dense $(n/2)$-lacings of the new shoe plus n (the total length of n horizontals of the original shoe) is the length of the corresponding dense-and-straight lacing of the original shoe. Therefore, the longest straight n-lacings are the dense-and-straight n-lacings that correspond to the longest dense $(n/2)$-lacings of the new shoe. By Theorem 6.2, the longest dense

$(n/2)$-lacing of the new shoe is the devil $(n/2)$-lacing. Finally, it is easy to show that the devil $(n/2)$-lacing of the new shoe corresponds to the straight devil n-lacings of the original shoe. This completes the proof for n even.

For odd n, the corresponding two-to-one correspondence and argument are generalizations of the correspondence of the straight devil 5-lacing and the dense 3-lacing shown in Figure 6.15. ☐

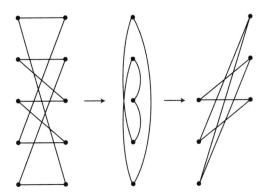

Figure 6.15. Turning a dense-and-straight n-lacing, n odd, that contracts to a longest one-column n-lacing into a dense $((n+1)/2)$-lacing (of a different linear shoe).

6.5 The n-Lacings of Longest Vertical Length

The *vertical length* of an n-lacing is the sum of the vertical lengths of its segments. Clearly, the shortest vertical length of an n-lacing is $2(n-1)$, the vertical length of any simple n-lacing. Also, it is easy to see that an n-lacing is simple if and only if its vertical length is $2(n-1)$. This means that since every one of the classes of n-lacings considered by us contains simple n-lacings, the n-lacings of shortest vertical length in every class are exactly the simple n-lacings contained in this class. For example, all the shortest n-lacings in every class also turn out to be of shortest vertical length.

If we fix the stretch of our mathematical shoe at 1 and let the distance between the two columns go to zero, then the length of a fixed lacing of this shoe will tend to its vertical length. An immediate consequence of this is the following theorem.

Theorem 6.8 (Longest for Long Shoes). *For long shoes, a longest n-lacing in one of our classes of lacings is of longest vertical length in this class.*

Note that not all the longest n-lacings in the different classes of n-lacings that we encountered in this chapter are of longest vertical length in these classes. An exception is the devil n-lacing for even n. It is the longest n-lacing for short shoes, but its vertical length turns out to be less than that of the angel n-lacing.

Let's calculate the vertical lengths of the angel n-lacings, the devil n-lacings, and the straight devil n-lacings. For even n, these vertical lengths are

$$|angel| = 2 \cdot 1 + 4[2 + 4 + 6 + \cdots + (n-2)] + 2(n-1)$$

$$= 2n + 8\left[1 + 2 + 3 + \cdots + \frac{n-2}{2}\right] = 2n + (n-2)n$$

$$= n^2,$$

$$|devil| = 4[2 + 4 + 6 + \cdots + (n-2)] + 2(n-1)$$

$$= 2n + 8\left[1 + 2 + 3 + \cdots + \frac{n-2}{2}\right] - 2$$

$$= n^2 - 2,$$

$$|st.devil| = 1 + 2[2 + 4 + 6 + \cdots + (n-2)] + (n-1)$$

$$= n + 4\left[1 + 2 + 3 + \cdots + \frac{n-2}{2}\right]$$

$$= \frac{n^2}{2}.$$

Similarly, we calculate that, for odd n, the vertical lengths of an angel n-lacing, a devil n-lacing, and a straight devil n-lacing are $n^2 - 1$, $n^2 - 1$, and $(n^2 - 1)/2$, respectively. Theorem 6.10, below, shows that the angel n-lacings are n-lacings of longest vertical length, that the devil n-lacings are dense n-lacings of longest vertical length, and that the straight devil n-lacings are straight n-lacings of longest vertical length.

To be able to describe the structure of the n-lacings of longest vertical length in the different classes of n-lacings, we need to introduce the *top* and *bottom parts* of our mathematical shoe. If n is even, the first $n/2$ pairs of eyelets are the top part, and the remaining $n/2$ pairs of eyelets the bottom part. If n is odd, the first $(n-1)/2$ and the last $(n-1)/2$ pairs of eyelets of our shoe are the top and bottom parts of our shoe, respectively. Note that in the odd case, the two middle eyelets are contained neither in the top part nor in the bottom part.

Remember that for a lacing to be dense just means that all its segments have endpoints in different columns. We call an n-lacing *dense in the vertical direction*, if: (1) for even n, all segments have endpoints in different parts; (2) for odd n, all segments have endpoints in different parts except for those segments ending in one of the two middle eyelets. Exactly half of these special segments have their second endpoint in the top par; the other half of these segments have their second endpoint in the bottom part. The angel n-lacings and the devil n-lacings for odd n are examples of n-lacings that are dense in the vertical direction.

We call an n-lacing, with n even, *close-to-dense in the vertical direction* if all its segments have endpoints in different parts except for one segment ending in one or both of the top two middle eyelets and a second segment ending in one or both of the bottom two middle eyelets. Note that this means that these two distinguished segments may be horizontals. The devil n-lacings, for n even, are examples of dense n-lacings that are close-to-dense in the vertical direction.

Let's call an n-lacing *dense-and-straight in the vertical direction* if it is straight and: (1) for even n, all nonhorizontal segments have endpoints in different parts; (2) for odd n, all nonhorizontal segments have endpoints in different parts except for those segments ending in one of the two middle eyelets. Exactly one of these two special segments has its second endpoint in the top part; the other has its second endpoint in the bottom part. The straight devil n-lacings are examples of n-lacings that are dense-and-straight in the vertical direction.

Theorem 6.9 (Characterization).
(General) The n-lacings of longest vertical length are exactly the n-lacings that are dense in the vertical direction.

(Straight) The straight n-lacings of longest vertical length are exactly the n-lacings that are dense-and-straight in the vertical direction.

(Dense) If n is odd, the dense n-lacings of longest vertical length are exactly the dense n-lacings that are dense in the vertical direction. If n is even, the dense n-lacings of longest vertical length are exactly the dense n-lacings that are close-to-dense in the vertical direction.

Theorem 6.8 implies that if we know a longest n-lacing in one of our classes of lacings for a long version of our mathematical shoe, then this n-lacing is also an n-lacing of longest vertical length in this class. For example, since we know that the devil n-lacing is the longest dense n-lacing, we conclude that the longest possible vertical length of a dense n-lacing is equal to the vertical length of the devil n-lacing. Using this argument, we can derive the longest vertical lengths of n-lacings in most of the classes of n-lacings under consideration. However, since we do not know whether the angel n-lacings are really the longest n-lacings for long shoes, we cannot derive the longest vertical length of n-lacings in this way. Therefore, we will use a different direct argument to prove the following result:

Theorem 6.10 (Longest Vertical Length). *The longest vertical lengths of n-lacings in the different classes of n-lacings under consideration are as summarized in the following Venn diagram:*

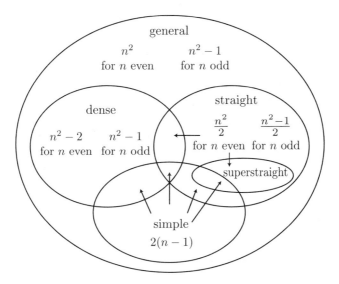

Figure 6.16. The longest vertical lengths of lacings in the different classes of n-lacings.

Note that, as an immediate consequence of these two theorems, for even n, no dense n-lacing is dense in the vertical direction.

For the proof of our two theorems we need the following lemma:

Lemma 6.11 (Even Vertical Length). *The vertical length of an n-lacing is an even number.*

Proof. If we identify the rows of an n-lacing with their indices, a journey around an n-lacing starting and ending at eyelet A_1 translates in a straightforward way into a walk on the numbers from 1 to n, starting and ending at 1, such that the vertical length of the n-lacing equals the number of steps of the walk. Now, clearly, on this walk, every step in the positive direction that we take has to be compensated by exactly one step in the negative direction for us to get back to 1. We conclude that every such walk is made up of an even number of steps.   □

Proof of Theorem 6.10. All simple n-lacings have vertical length $2(n-1)$. Therefore, the longest vertical length of a simple n-lacing is also $2(n - 1)$. This takes care of the simple n-lacings and the four simple-and-... classes of n-lacings.

The vertical length of a straight n-lacing is the length of its contraction (see Chapter 2), and for every one-column n-lacing there is a dense-and-straight n-lacing that contracts to this one-column n-lacing; see Theorem 2.4. Therefore, the longest vertical length of a straight n-lacing or a dense-and-straight n-lacing equals the maximal length of a one-column n-lacing, that is, the length given in the diagram; see also Theorem 2.3.

As we have seen, the devil n-lacings and the angel n-lacings provide examples of dense and general n-lacings whose vertical lengths equal those listed in the diagram under "general" and "dense". This means that to finish the proof of this theorem it remains to show that the numbers under "general" and "dense" are upper bounds for the respective longest vertical lengths.

Let's choose the stretch of our mathematical shoe to be 1 and, as we did in the special case of straight n-lacings in Chapter 2, contract one of our lacings in the horizontal direction until the two columns of eyelets coincide. On the one hand, this turns our mathematical shoe into a *contracted n-shoe* consisting of n eyelets, which we will number 1 to n, from top to bottom. On the other hand, the lacing turns into an *n-contraction*. The segments of this n-contraction are the contractions of the individual segments of the n-lacing that we started with. Since horizontals contract to points, we don't count these contractions among the segments of the n-contraction. Note that the n-contraction of a straight n-lacing is its contraction, as we defined it in Chapter 2. Furthermore, an n-contraction has the following properties:

(C1) It consists of at most $2n$ segments whose endpoints are different eyelets of the contracted n-shoe.

(C2) Every eyelet of the contracted n-shoe is endpoint of at most four segments.

We call any set of segments that satisfies these two conditions an *exploded n-contraction*.

In the following, we will assume that terms like the length of an (exploded) n-contraction or the top and bottom parts of a contracted shoe are defined as for lacings and mathematical shoes.

Clearly, the length of an exploded n-contraction that corresponds to an n-lacing equals the vertical length of the n-lacing. We now show that the maximal length of an exploded n-contraction is n^2 for even n and $n^2 - 1$ for odd n, which then implies that the longest vertical length of an n-lacing is as claimed. The proof is similar to our proof that $n^2/2$ and $(n^2 - 1)/2$ are the maximal lengths of one-column n-lacings for even and odd n, respectively; see Chapter 2.

Let M be an exploded n-contraction of maximal length. An eyelet E has k_E segments of M ending in it, where, because of property C2, the number k_E is 0, 1, 2, 3, or 4. The eyelet E is k-*full* if $k = k_E$. Because of property C1, the sum of the k_Es equals two times the number of segments in M. In particular, all eyelets are 4-full if and only if there are $2n$ segments in M. If there were two different eyelets that are not 4-full, then there are less than $2n$ segments, and we could lengthen M into a new exploded n-contraction by inserting an extra segment with endpoints in these two eyelets. Since this is not possible, we conclude

(C3) There is at most one eyelet that is not 4-full. If there is such an eyelet, then it is either 0-full or 2-full. In the first case M contains $2n - 2$ segments; in the second case it contains $2n - 1$ segments.

Let's first assume that there is an eyelet E that is not 4-full. If there was a segment in M both of whose endpoints are situated above E or both below, then we could lengthen this segment by moving the endpoint that is closer to E until it coincides with E. Since this would also yield a longer exploded n-contraction, we conclude that a segment of M either ends in the eyelet E or one of its endpoints is above and the other below E. This means that the special eyelet E separates the endpoints of all except for k_E segments. If $k_E = 0$, then the number of segments equals both the number of eyelets below and the number of eyelets above the special eyelet. This is impossible if n is even, and, if n is odd, it is clear that E has to be the middle eyelet. If $k_E = 2$, then there are two segments S and T ending in it. Let's assume that there are m eyelets above E. Then, if both S and T end above E, we conclude that the number of segments in M equals $4m$, the number of segments ending in the first m eyelets. This number is also equal to $4(n - m - 1) + 2$, the number of segments ending in the eyelets below E plus the two segments ending in E. However, the equation $4m = 4(n-m-1)+2$ clearly does not have any integer solutions. Now, if S ends above E and T below, then we conclude for the number of segments $2n - 1 = 4m + 1$ and $2n - 1 = 4(n - m - 1) + 1$. Since both equations reduce to $n = 2m + 1$, we conclude that E is the middle eyelet. This gives

(C4) If there is an eyelet E that is not 4-full, then n is odd and E is the middle eyelet. The two endpoints of segments that do not end in E are contained in different parts. If the eyelet E is 2-full, then the endpoints different from E of the two segments ending in E are contained in different parts.

Let's deal with the case that all eyelets are 4-full. If n is even, let's assume that there is a segment both of whose endpoints are in the top part. Then there is also a segment both of whose endpoints are contained in the bottom part (otherwise there would be at least $4 \cdot n/2 + 1 = 2n + 1$ segments in M, which is impossible.) But then we modify our maximal exploded n-contraction M as in Figure 6.17 into a longer exploded n-contraction, which is impossible. We conclude that given a segment of M, one of its endpoints is contained in the top part and the other in the bottom part.

Figure 6.17. Lengthening an exploded n-contraction.

Similarly, for odd n, if there was a segment in M both of whose endpoints are in the top part, then all segments that have an endpoint in the bottom part or the middle eyelet would have their second endpoints in the top part. Otherwise, M could be lengthened as in Figure 6.17, which is impossible. However, this would also mean that there are at least $4(n-1)/2 + 4 + 1 = 2n + 3$ segments in M, which is not the case. We conclude that there is no segment in M both of whose endpoints are contained in either the top part or the bottom part. There are four segments ending in the middle eyelet. As in the case of a middle 2-full eyelet, we conclude that for this 4-full middle eyelet two of the segments ending in it are "going up" and two are "going down".

(C5) If all eyelets are 4-full and n is odd, then two of the segments ending in the middle eyelet have their second endpoints in the top part and the other two in the bottom part. The two endpoints of segments that do not end in E are contained in different parts. If all eyelets are 4-full and n is even, then the endpoints of all segments are contained in different parts.

Let n be even. Then we know that all eyelets are 4-full. If M does not contain four segments of maximal possible length $n - 1$, pick a segment T that ends in eyelet 1 and a segment S that ends in eyelet n such that the length of both S and T is less than $n - 1$. Then, because of property C5, both segments are interlaced as the two segments shown on the left side of Figure 6.18. We now modify M by replacing T and S by the two segments T' and S', as shown in this diagram. This yields an exploded n-contraction that has one more segment of length $n-1$ than M and has the same length as M; that is, the new exploded n-contraction also has maximal length. We continue modifying M like this until it contains four segments of length $n - 1$. These are then exactly the segments ending in eyelets 1 and n. Using the same method, we now modify M until it also contains four segments of length $n - 3$, which are then exactly the segments ending in eyelets 2 and $n - 1$. Continuing like this, we can turn M into an exploded n-contraction of maximal length consisting of four segments of length $n - 1$, four of length $n - 3$,..., and four of length 1. Hence the maximal length of an exploded n-contraction for even n is

$$4(1 + 3 + 5 + \cdots + (n - 1)) = n^2.$$

The case that n is odd can be reduced to the even case as follows. Remove the middle eyelet and combine segments ending in it in pairs into new segments whose two endpoints are contained in different parts. This is possible, since in all possible cases among the segments ending in the middle eyelet there is the same number

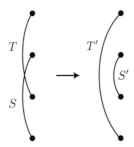

Figure 6.18. Modifying an exploded n-contraction such that its length does not change.

of segments that have the second endpoint in the top part as there are segments with second endpoint in the bottom part. Now, move the bottom part up by one unit, shortening all segments as you go. This gives an exploded $(n-1)$-contraction, which is easily seen to be of maximal length. Working backwards, we find that if the middle eyelet is 4-full, the maximal length of an exploded n-contraction with n odd is

$$(n-1)^2 + 2(n-1) = n^2 - 1.$$

Note that for this argument it does not matter whether the middle eyelet is 4-full, 2-full, or 0-full. This confirms that the maximal vertical length of a general n-lacing is n^2 in the even case and $n^2 - 1$ in the odd case.

Since, for odd n, the devil n-lacing has vertical length $n^2 - 1$, we know that the longest vertical length of a dense n-lacing is $n^2 - 1$. On the other hand, for even n, the devil n-lacing has only vertical length $n^2 - 2$. Since, by Lemma 6.11, the vertical length of an n-lacing is even, the longest vertical length of a dense n-lacing for even n is either n^2 or $n^2 - 2$. However, it is easy to see that it cannot be n^2. Just assume that there was a dense n-lacing with vertical length n^2. Then it is also dense in the vertical direction. This means that when we go for a round trip around this lacing starting at eyelet A_1, we move to the bottom part and the right column on step one, the top part and the left column on step two, the bottom part and the right column on step three, and so on. This means that we never visit the top part of the right column or the bottom part of the left column, which is impossible. This confirms that the maximal vertical length of a dense n-lacing is $n^2 - 2$ in the even case and completes the proof of this theorem. □

Proof of Theorem 6.9. We first show that an n-lacing is of longest vertical length if and only if it is dense in the vertical direction. Continuing the arguments in the previous proof, we immediately conclude that every n-lacing of longest vertical length is dense in the vertical direction.

Conversely, if an n-lacing is dense in the vertical direction and n is even, then it is clear that the exploded n-contraction associated with this n-lacing can be turned into an exploded n-contraction of the same length consisting of four segments of length $n-1$, four segments of length $n-3$, ..., and four segments of length 1. The method for doing this is the same as the one that we used in the previous proof. Calculating the vertical length of the new exploded n-contraction, we conclude that our n-lacing is of longest vertical length. If an n-lacing is dense in the vertical direction and n is odd, we can again use the same arguments as in the previous proof to convince ourselves that we are dealing with an n-lacing of longest vertical length. This completes the proof of the first part of this theorem.

The characterization of the straight n-lacings of longest vertical length follows in a similar fashion from the structure of the one-column lacings of maximal length to which these n-lacings contract.

We now derive the structure of the dense n-lacings of longest vertical length for even n. A dense lacing consists of segments of positive, negative or zero slope. The segments of zero slope are just the horizontals. Since dense lacings do not contain verticals, they do not contain segments of infinite slope. It is easy to see that a connected component of a dense lacing consisting of segments of positive slope only will never contain eyelets A_1 and B_n. A connected component consisting of segments of nonnegative slope can contain all eyelets. However, if it contains eyelet A_1, then this connected component necessarily contains the top horizontal and has A_1 as one of its ends. Similarly for B_n and negative-slope-only components. We conclude that dense lacings contain both segments of positive and negative slope. As we trace a dense lacing, a change from positive to negative slope segments has to happen via a *switch*, that is, two consecutive segments PQ and QR of different sign (positive, negative, or zero); see Figure 6.19.

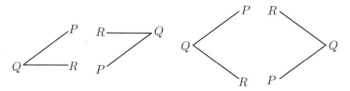

Figure 6.19. The four possible kinds of switches PQR from positive to negative slope in a dense lacing.

Since a lacing is a closed loop, it has to contain at least two switches. Furthermore, there are two switches that do not overlap, that is, are disjoint. To see this, consider the comprehensive list of overlapping switches in Figure 6.20. Then it is easy to see that such a configuration can be completed to a dense lacing only via an extra switch. This new switch and one of the switches in the configuration we started with are then definitely disjoint.

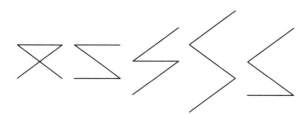

Figure 6.20. The five different types of two switches that overlap in one segment (up to reflections in horizontals and verticals).

Now, select two disjoint switches S and T and contract the dense lacing as in the previous proof, but instead of contracting the segments that S and T consist of individually, we contract S and T to one segment each; see Figure 6.21 for an example. What we end up with is a modified n-contraction with the following properties:

(MC1) It contains at most $2n - 2$ segments.
(MC2) Every eyelet is the endpoint of at most 4 segments.

An *exploded modified n-contraction* is a set of segments of a contracted n-shoe with these properties.

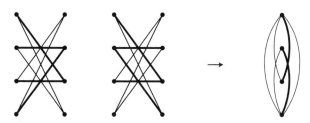

Figure 6.21. Up to symmetries of the underlying shoe, there are two ways to choose two disjoint switches (drawn thick) in a devil 4-lacing. The corresponding modified 4-contractions happen to coincide.

Let M be an exploded modified n-contraction of maximal length. Remember that n is even. If M consisted of less than $2n - 2$ segments, then $4n - 2(2n - 3) = 6$, which would mean that at least two eyelets are not full. Then we could lengthen M to a new exploded modified n-contraction by adding a segment that ends in two eyelets that are not full. We conclude that M contains the maximum possible number of segments, $2n - 2$. Since $4n - 2(2n - 2) = 4$, the possible configurations of eyelets that are not full are as follows:

1. One eyelet 0-full.
2. One eyelet 1-full and one eyelet 3-full.
3. Two eyelets 2-full.
4. One eyelet 2-full and two eyelets 3-full.
5. Four eyelets 3-full.

We show that configurations (1), (4), and (5) are not possible. If configuration (1) were possible, we could conclude, as in the previous proof, that every segment has one endpoint above and one below the distinguished eyelets. This implies that there are the same number of eyelets above and below this eyelets. But this is only possible if n is odd, which is not the case. In both configurations (4) and (5) there are three eyelets that are not full. From the middle one of the three a segment is going up (or down). Then we can stretch this segment down (or up) to create a longer exploded modified n-contraction. Since this is not possible, we conclude that configurations (4) and (5) are not present in M. In the remaining two configurations there are exactly two eyelets with holes. If these eyelets were not adjacent, we could stretch one of the segments ending in an eyelet in between these two eyelets to create a longer exploded modified n-contraction. Since this is not possible, we conclude that the two eyelets that are not full are adjacent.

We now show that configuration (2) is not possible as well. With similar arguments as above, we can convince ourselves that if configuration (2) were present in M, then the segments ending in the two eyelets that are not full would be distributed as in one of the two diagrams in Figure 6.22.

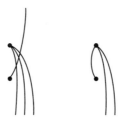

Figure 6.22.

All other segments would have one endpoint above and one below the distinguished two eyelets each. Let's consider the left diagram and assume that there are m eyelets above the distinguished two. Comparing the numbers of endpoints of the segments that do not end in the distinguished eyelets above and below the distinguished eyelets, we get $4m-1 = 4(n-m-2)-3$. This is equivalent to $2(n-2m) = 5$, an equation that does not have any integer solutions. Similarly, the right diagram corresponds to the equation $4m = 4(n - m - 2) - 2$, which is also equivalent to the equation $2(n - 2m) = 5$. We conclude that configuration (2) cannot occur in M.

This means that in M configuration (3) is the only possible configuration of eyelets that are not full. In addition, arguing as above, we can convince ourselves that the segments ending in the eyelets that are not full would be distributed as in one of the three diagrams in Figure 6.23 and that these distinguished eyelets are indeed the middle two eyelets.

Figure 6.23.

Using the same approach that we applied to evaluate the length of a maximal collapsed n-lacing, it is easy to show that in all three cases the length of M, an exploded modified n-contraction of maximal length, is $n^2 - 2$. Since this is also the maximal vertical length of a dense n-lacing, a modified n-contraction of such a dense n-lacing is also of maximal length and, therefore, corresponds to one of these three diagrams.

Let N be a dense n-lacing of maximal vertical length. Then it is clear from what we have deduced so far that: (1) any segment of N that does not end in the middle two rows has one of its endpoints in the top part and one in the bottom part; (2) whenever you choose two disjoint switches in N, one of the switches has its joint (the point at which the two segments of the switch meet) in the top middle row, and the other switch has its joint in the bottom middle row.

It remains to figure out what N can possibly look like near the middle four eyelets. In Figure 6.24 we list the possible four essentially different switches in N, with joints in the middle two rows, whose contractions can contribute to the three configurations in Figure 6.23.

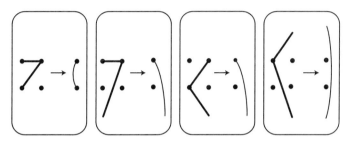

Figure 6.24.

In Figure 6.25 we list all possible disjoint pairs of such switches such that the joint of one is contained in the top middle row and the joint of the second in the bottom middle eyelet.

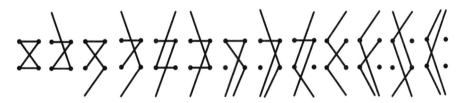

Figure 6.25.

In Figure 6.26 we extend these possibilities to a complete list of the different possible local configurations for N around the middle four eyelets. Inspection of these different possible configurations yields that N is close-to-dense in the vertical direction.

Finally, it is easy to see that every dense n-lacing that is close-to-dense in the vertical direction is also of maximal vertical length. This completes the proof of this theorem. □

6.6 Notes

The n-lacings of longest vertical length are in some sense the opposite of simple n-lacings and do form another very natural class of n-lacings besides the ones considered by us. It would be interesting to count the number of n-lacings of longest vertical length in the different classes of n-lacings and to find the shortest among them.

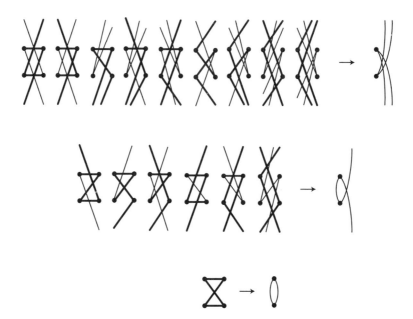

Figure 6.26.

Figure 6.27. A real devil 4-lacing and a real angel 4-lacing.

7

The Strongest Lacings

When you pull on the ends of a shoelace, a lacing acts like a pulley. In the following, we will determine which of the n-lacings in the different classes of n-lacings under consideration are the strongest pulleys.

When a shoelace is tied properly, we may assume that, ideally, the tension along the shoelace is a positive constant T. Then this tension translates into a tension T_h of the pulley in the horizontal direction, that is, the direction in which the two sides of the shoe are being pulled together. This tension T_h is the sum over all horizontal components of T along the different segments of the lacing. Clearly, if we are dealing with a vertical segment, then this horizontal component is 0; and if we are dealing with a diagonal segment A_iB_j, as in Figure 7.1, then we calculate for the horizontal component $T_{i,j}$ along this segment $\frac{T_{i,j}}{T} = \frac{1}{|A_iB_j|}$, that is,

$$T_{i,j} = \frac{T}{|A_iB_j|}.$$

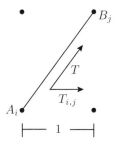

Figure 7.1. The horizontal component $T_{i,j}$ of the tension T along the segment A_iB_j is $T/|A_iB_j|$.

Without loss of generality, we may set $T = 1$. We call the value of T_h associated with an n-lacing l its *pulley sum*. Clearly,

$$T_h = \sum_{d \in \{\text{diagonals of } l\}} \frac{1}{|d|}.$$

A *strongest n-lacing* in a set of n-lacings is a lacing which has a maximal pulley sum. The results of this chapter are summarized in the following simple Venn diagram:

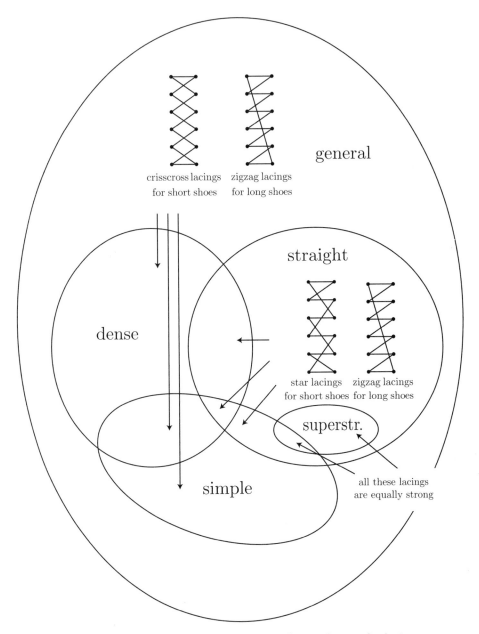

Figure 7.2. The strongest lacings in the different classes of n-lacings.

Note that the star, crisscross and zigzag n-lacings that are the strongest lacings in the general and straight categories are also dense and simple. Therefore, the strongest representatives in most of the other classes are also of these types. For example, the strongest simple n-lacings are just the strongest n-lacings overall.

7.1 Statement of Results

In everything that follows, it is important to remember that h denotes the stretch of a mathematical shoe. Here are the two main results of this chapter.

Theorem 7.1 (Strongest). *The unique dense 2-lacing is the strongest 2-lacing. Let $n > 2$ and let $C(n,h)$ and $Z(n,h)$ be the pulley sums of the crisscross n-lacing and the zigzag n-lacings, respectively. Then there is exactly one $h_n > 0$ such that $C(n, h_n) = Z(n, h_n)$. Furthermore, depending on h, the strongest n-lacings are:*

- *the crisscross n-lacing for $h < h_n$;*
- *the crisscross n-lacing and the zigzag n-lacings for $h = h_n$;*
- *the zigzag n-lacings for $h > h_n$.*

I find this result very satisfactory, as it characterizes the two most commonly used types of lacings as the strongest lacings overall. However, note that if we choose h much smaller than h_n, the zigzag n-lacings will, in general, be far from being the second strongest n-lacings. Similarly, for h much larger than h_n, the crisscross n-lacing will be far from being the second strongest n-lacing. In fact, we will see in the next chapter that in the class of dense-and-simple n-lacings the zigzag n-lacings are the *weakest* lacings for small h and the crisscross n-lacings are the weakest ones for large h. So, in this class, in terms of strength, our two favorite lacings are as far apart as possible for both long and short shoes. See the notes at the end of the next chapter for more details about this remark.

Table 7.1 lists the approximate values of h_n for small n. It is interesting to note that for many real shoes with n pairs of eyelets, the ratio of the distance between adjacent rows of eyelets and the distance between the columns of eyelets is very close to h_n. This means that no matter whether you prefer to lace zigzag or crisscross, you get close to maximizing the horizontal tension when you pull on the two ends of one of your shoelaces.

Table 7.1. The approximate value of the stretch h_n for which the crisscross n-lacing is as strong as the zigzag n-lacings.

n	3	4	5	6	7	8	9	10
h_n	0.9029	0.7412	0.6450	0.5794	0.5309	0.4931	0.4625	0.4372

Theorem 7.2 (Strongest Straight). *The zigzag 2- and 3-lacings, which are the same as the star 2- and 3-lacings, respectively, are the strongest straight 2- and 3-lacings. Let $n > 3$ and let $S(n,h)$ and $Z(n,h)$ be the pulley sums of the star n-lacings and the zigzag n-lacings, respectively. Then there is exactly one $h_n > 0$ such that $S(n, h_n) = Z(n, h_n)$. Furthermore, depending on h, the strongest straight n-lacings are:*

- *the star n-lacings for $h < h_n$;*
- *the star n-lacings and the zigzag n-lacings for $h = h_n$;*
- *the zigzag n-lacings for $h > h_n$.*

We will see in the next chapter that in the class of dense-and-simple-and-straight n-lacings the zigzag n-lacings are the *weakest* lacings for small h and the star n-lacings are the weakest ones for large h. So, in this class, in terms of strength, these two types of lacings are as far apart as possible for both long and short shoes.

Table 7.2 lists the approximate values of h_n for small n.

Table 7.2. The approximate value of the stretch h_n for which the star n-lacings are as strong as the zigzag n-lacings.

n	4	5	6	7	8	9	10
h_n	0.3720	0.3312	0.3019	0.2794	0.2615	0.2468	0.2344

On closer inspection, Theorems 7.1 and 7.2 are rather surprising results. It is NOT surprising that the crisscross n-lacing is the strongest n-lacing for small values of h and that the zigzag n-lacings are strongest for large values of h. What is surprising is the fact that there are no other n-lacings that are stronger than both the crisscross and zigzag n-lacings for "intermediate" values of h. After all, the crisscross and zigzag lacings are not really what you would call close relatives, as it is not possible to turn one into the other by making small changes in the lacings. Also, there are a number of "hybrid" lacings that start out as crisscross lacings and continue as zigzag lacings; see Figure 7.3. It would be reasonable to expect that some of these hybrid lacings could take over as the strongest lacings for intermediate values of h. However, this turns out not to be the case.

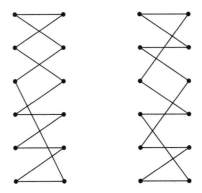

Figure 7.3. On the left, a hybrid 6-lacing that is half crisscross half zigzag lacing, and, on the right, a hybrid 6-lacing that is half star and half zigzag lacing.

7.2 The Strongest n-Lacings

Our strategies for proving that the classical n-lacings are the strongest n-lacings overall and that the star and the zigzag n-lacings are the strongest straight n-lacings are very similar. Here is a brief outline of our strategy for deriving the first of these two results. First, we introduce exploded n-lacings. Every n-lacing gives rise to an exploded n-lacing. We define a pulley sum for exploded n-lacings such that the pulley sum for an n-lacing and its associated exploded n-lacing coincide. Using some very powerful strengthening rules, we prove that the strongest exploded n-lacings are the two exploded classical n-lacings. Since the only n-lacings that give rise to the exploded classical n-lacings are the classical n-lacings, it follows that the strongest n-lacings are the classical n-lacings. Here are the details.

The collection V of the vertical lengths of diagonals of an n-lacing has the following properties:

(E1) V consists of at most $2n$ nonnegative integers less than or equal to $n - 1$.
(E2) V contains at most n 0s.
(E3) If V contains $2n$ elements, then the sum of the integers in V is greater than or equal to $2(n - 1)$.

Here property E1 just says that every n-lacing contains at most $2n$ diagonals and that the maximal vertical length of a diagonal is $n - 1$. Property E2 says that an n-lacing can have at most n horizontals. Finally, starting from eyelet A_1, we travel along the lacing to eyelet A_n and back. In the course of this journey, we cover a total vertical distance that is greater than or equal to $2(n - 1)$. This means that if all segments of an n-lacing are diagonals, then the sum of the vertical lengths of these diagonals is greater than or equal to $2(n - 1)$. This is property E3.

In general, we call a collection of at most $2n$ integers an *exploded n-lacing* if it has properties E1, E2, and E3 above.

We define the *pulley sum* of a collection of integers V to be the sum

$$\sum_{s \in V} \frac{1}{\sqrt{1 + (hs)^2}}.$$

Clearly, the pulley sum of an exploded n-lacing that arises from an n-lacing is the same as the pulley sum of this n-lacing.

We define three strengthening rules that allow us to turn most exploded n-lacings into stronger exploded n-lacings. Let V be an exploded n-lacing, $n > 2$.

Strengthening Rule 1. Assume that V contains $m < 2n$ elements. If $m < n$, then we add $n - m$ 0s, $n - 1$ 1s and one $n - 1$ to V to get a new stronger exploded n-lacing V' with $2n$ elements. If $n \le m < 2n$, then we add $2n - m - 1$ 1s and one $n - 1$ to V to get a new stronger exploded n-lacing V' with $2n$ elements. (In both cases it is obvious that V' has properties E1 and E2 and that it is stronger than V. Furthermore, since, by E2, the original set contains at most n 0s, the sum of the $2n$ elements of V' is clearly greater than or equal to $(n - 1) \cdot 1 + 1 \cdot (n - 1) = 2(n - 1)$. Hence V' also has property E3 and is therefore really a new exploded n-lacing.)

Strengthening Rule 2. Assume that V contains exactly $2n$ elements and that the sum of its elements is greater than $2(n - 1)$. If V contains less than n 0s, then we can replace one of its nonzero elements v by $v - 1$ to arrive at a stronger exploded n-lacing. Otherwise, V contains at least one element v greater than 1. By replacing this element by $v - 1$, we arrive at a new stronger exploded n-lacing.

Strengthening Rule 3. Consider any two elements v, v' of V whose sum is greater than or equal to 2. Assume that there are nonnegative integers w and w' less or equal to $n - 1$ such that the following conditions are satisfied:

- The sum of w and w' is equal to that of v and v'.
- The pulley sum of w and w' is greater than that of v and v'.
- If V contains n 0s and both v and v' are nonzero, then neither w nor w' is 0.

Then replacing v and v' in V by w and w' yields an exploded n-lacing that is stronger than V and the sum of whose elements is the same as that for V. (Note that the last condition makes sure that the new collection of integers has property E2.)

Not every exploded n-lacing arises from an n-lacing, and there may be a number of very different n-lacings that give rise to the same exploded n-lacing. However, there are exploded n-lacings that correspond to very few n-lacings. In particular, the *exploded crisscross n-lacing*, consisting of two 0s and $2(n-1)$ 1s, and the *exploded zigzag n-lacing*, consisting of n 0s, $n-1$ 1s, and one $n-1$, are of this type.

Lemma 7.3. *The only n-lacing corresponding to the exploded crisscross n-lacing is the crisscross n-lacing. The only two n-lacings corresponding to the exploded zigzag n-lacing are the two zigzag n-lacings.*

We call an exploded n-lacing *reducible* or *irreducible* depending on whether or not it can be strengthened using one of our strengthening rules. Now, for any possible choices of n and h, the strongest exploded n-lacings are clearly irreducible. In fact, we will prove the following crucial lemma:

Lemma 7.4. *For any of the possible choices for $n > 2$ and $h > 0$, let V be an exploded n-lacing that is not the exploded crisscross n-lacing or the exploded zigzag n-lacing. Then V is reducible.*

Finally, Lemmas 7.3 and 7.4, together with the following two lemmas, combine into a proof of Theorem 7.1 in a straightforward manner.

Lemma 7.5. *The unique dense 2-lacing is the strongest 2-lacing for all $h > 0$.*

Lemma 7.6. *Let $n > 2$, $h > 0$, and let $C(n,h)$ and $Z(n,h)$ be the pulley sums of the crisscross n-lacing and the zigzag n-lacings, respectively. Then there is exactly one $h_n > 0$ such that $C(n, h_n) = Z(n, h_n)$. Furthermore,*

$$C(n,h) > Z(n,h) \text{ for } h < h_n,$$

and

$$C(n,h) < Z(n,h) \text{ for } h_n < h.$$

7.3 The Strongest Straight n-Lacings

To prove Theorem 7.2, we proceed as follows: A *straight exploded n-lacing* is an exploded n-lacing that contains n 0s. Clearly, every straight n-lacing gives rise to a straight exploded n-lacing.

We call a straight exploded n-lacing *straight-reducible* or *straight-irreducible* depending on whether or not it can be strengthened into another straight exploded n-lacing using one of our straightening rules. Now, for any possible choices of n and h, the strongest straight exploded n-lacings are clearly straight-irreducible. In fact, we will prove the following crucial lemma:

Lemma 7.7. *For any of the possible choices of $n > 3$ and $h > 0$, let V be a straight exploded n-lacing that is not the exploded star n-lacing or the exploded zigzag n-lacing. Then V is straight-reducible.*

Lemma 4.5 and Lemma 7.3 guarantee that the only straight n-lacings that give rise to the exploded star n-lacing and the exploded zigzag n-lacing are the star n-lacings and zigzag n-lacings, respectively. Now, this fact, Lemma 7.7 and the following two lemmas combine into a proof of Theorem 7.2 in a straightforward manner.

Lemma 7.8. *The unique dense 2-lacing is the strongest straight 2-lacing for all $h > 0$. The zigzag 3-lacing is the strongest straight 3-lacing for all $h > 0$.*

Lemma 7.9. *Let $n > 3$, $h > 0$, and let $S(n, h)$ and $Z(n, h)$ be the pulley sums of the star n-lacings and the zigzag n-lacings, respectively. Then there is exactly one $h_n > 0$ such that $S(n, h_n) = Z(n, h_n)$. Furthermore,*

$$S(n, h) > Z(n, h) \text{ for } h < h_n,$$

and

$$S(n, h) < Z(n, h) \text{ for } h_n < h.$$

7.4 Proofs of the Lemmas

We leave the proofs of Lemmas 7.3, 7.5, and 7.8 as exercises and now give the proofs for Lemmas 7.4, 7.6, 7.7, and 7.9.

Proof of Lemma 7.4. We want to show that, for fixed $n > 2$ and $h > 0$, any exploded n-lacing different from the exploded crisscross n-lacing and the exploded zigzag n-lacing is reducible.

Let V be an irreducible exploded n-lacing. As a consequence of strengthening rule 1, we conclude that V has $2n$ elements.

As a consequence of strengthening rule 2, we conclude that

(E4) the sum of the elements of V is $2(n - 1)$.

Given that the sum of the $2n$ elements of V is $2n - 2$ and there is a maximum of n 0s, we draw the following conclusions:

(E5) V contains at least two 0s;
(E6) If V contains exactly two 0s, then it is the exploded crisscross n-lacing.
(E7) If V contains only 0s and 1s, then it is the exploded crisscross n-lacing.

Further, if V contains an $n - 1$, then the sum of the remaining $2n - 1$ elements in V is $n - 1$. Since there are at most n 0s in V and

$$0n + 1(n - 1) = n - 1,$$

we conclude that V consists of n 0s, $n - 1$ 1s, and the single $n - 1$; that is,

(E8) if V contains an $n - 1$, then it is the exploded zigzag n-lacing.

With an argument similar to the one we just used, we conclude that

(E9) V does contain 1s.

In the case $n = 3$, properties E7 and E8 together imply that V is either the exploded crisscross or zigzag n-lacing. Therefore, we may fix $n > 3$ in everything that follows.

We now proceed to investigate what strengthening rule 3 implies for irreducible exploded n-lacings such as V.

Let $k > 0$, and consider the function

$$P_k : [0, k] \to \mathbf{R} : x \mapsto \frac{1}{\sqrt{1 + x^2}} + \frac{1}{\sqrt{1 + (k - x)^2}}.$$

Figure 7.4 shows the graph of this function for twenty equally spaced choices of k. Note that if v and v' are two elements of V, then the pulley sum of these two elements is $P_{(v+v')h}(vh) = P_{(v+v')h}(v'h)$.

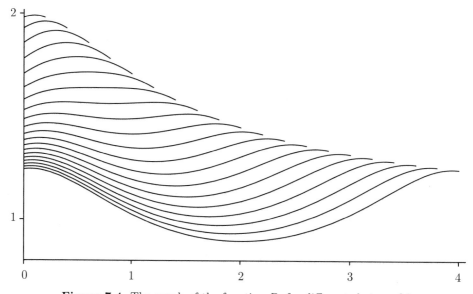

Figure 7.4. The graph of the function P_k for different choices of k.

In the following, we list some important properties of the function P_k.

(P1) P_k has a local extremum at $k/2$ and is symmetric about the vertical through this extremum.
(P2) Depending on the value of k, the function P_k is one of two types. If $k \leq \sqrt{2}$, then it is of *type 1*. In this case, P_k is strictly increasing from 0 to $k/2$. This means that the function has an absolute maximum at $k/2$. If $k > \sqrt{2}$, then the function P_k is of *type 2*. In this case, there is a value $x_k \in]0, k/2[$ such that P_k is strictly increasing in the interval $[0, x_k]$ and strictly decreasing in the interval $[x_k, k/2]$. We verify that P_k really has these properties in

Lemma 7.10. (Note that $P_k(x) = g_{k/2}(x - k/2)$, where g_j is the function considered in Lemma 7.10. This means that the properties of g_j that we verify there translate in a straightforward manner into the properties of P_k above.)

(P3) If $k_0 \geq 2h > 0$ and $P_{k_0}(0) \geq P_{k_0}(h)$, then $P_k(0) > P_k(h)$ for all $k > k_0$; see Lemma 7.11 below.

(P4) If $k_0 \geq 4h > 0$ and $P_{k_0}(h) \geq P_{k_0}(2h)$, then $P_k(h) > P_k(2h)$ for all $k > k_0$; see Lemma 7.11 below.

We are now able to finish the proof of this lemma. Let's assume that our irreducible exploded n-lacing V is different from the exploded crisscross and zigzag n-lacings. Property E8 implies that

(E10) all elements of V are less than $n - 1$.

Because of property E7, we know that

(E11) V contains at least one element that is greater than 1. Let e be a smallest element of V greater than 1.

If the pulley sum of a 0 and an e is less than the pulley sum of a 1 and an $e - 1$, that is,

$$P_{eh}(0) < P_{eh}(h),$$

then we can replace one of the 0s in V (see property E5) and an e by a 1 and an $e - 1$ to arrive at a stronger exploded n-lacing. Since V is irreducible, this is not possible. Therefore, we conclude that

(E12) $P_{eh}(0) \geq P_{eh}(h)$.

Hence, as a consequence of property P3, we find that

$$P_{(e+1)h}(0) > P_{(e+1)h}(h).$$

If V contains less than n 0s, this inequality tells us that we can strengthen V by replacing a 1 (see property E9) and an e by a 0 and an $e + 1$. Since this would contradict the fact that V is irreducible, we conclude that

(E13) V contains n 0s.

This means that the n elements of V that are not 0s have to add up to $2n - 2$. Remember that $e \leq n - 2$. Consequently, apart from e, V contains $n - 1$ nonzero elements whose sum is at least n. Hence, at least one of these elements, let's call it f, is greater than 1, and since e is supposed to be minimal among all elements of V having this property, we also know that $e \leq f$. Again, as a consequence of properties E12 and P3, we find that

$$P_{(e+f)h}(0) > P_{(e+f)h}(h).$$

For this to be the case, the function $P_{(e+f)h}$ has to be of type 2, and the value $x_{(e+f)h}$ (see property P2) at which $P_{(e+f)h}$ is maximal has to be less than h; see also Figure 7.5.

Then, because $P_{(e+f)h}$ is strictly decreasing from $x_{(e+f)h}$ to $(e + f)h/2$, and

$$x_{(e+f)h} < h < eh \leq (e + f)h/2,$$

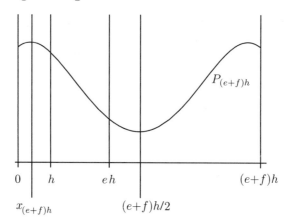

$x_{(e+f)h}$

$(e+f)h/2$

Figure 7.5.

we also conclude that

$$P_{(e+f)h}((e-1)h) > P_{(e+f)h}(eh).$$

Hence, we can strengthen V by replacing e and f by $e-1$ and $f+1$. This is the contradiction we are after, and we conclude that the only possible candidates for irreducible exploded n-lacings are the exploded crisscross and zigzag n-lacings. □

Proof of Lemma 7.7. We want to show that for fixed $n > 2$ and $h > 0$, any straight exploded n-lacing different from the exploded star n-lacing and the exploded zigzag n-lacing is straight-reducible.

Let V be a straight-irreducible straight exploded n-lacing. As in the proof of Lemma 7.4 (see also Lemma 4.5.2), we conclude that

(F1) V has $2n$ elements that sum up to $2(n-1)$.
(F2) V contains a 1.
(F3) If V contains only 0s, 1s, and 2s, then it is the exploded star n-lacing.
(F4) If V contains an $n-1$, then it is the exploded zigzag n-lacing.

In the cases $n = 3$ and $n = 4$, properties F3 and F4 together imply that V is either the exploded star or zigzag n-lacing. Therefore, we may fix $n > 4$ in everything that follows.

We assume that V is a straight-irreducible straight exploded n-lacing that is different from the exploded star and zigzag n-lacings.

First, we note that V contains at least two elements greater than 1. If this was not the case, V would contain n 0s, $n-1$ 1s and one element which, as a consequence of F4 and our assumption, is less than or equal to $n-2$. This means that the sum of its elements would be less than or equal to $n-1+n-2 = 2n-3$, which would contradict property F1. Because of property F3, we also know that there is at least one element greater than 2. Therefore, we choose two elements of V such that

(F5) e is a smallest element of V greater than 1 and f is another element of V that
    is a largest element greater than 2. In particular $1 < e \leq f < n-1$.

If

$$P_{(1+f)h}(h) < P_{(1+f)h}(2h),$$

then we can replace one of the 1s (see property F2) and an f by a 2 and an $f-1$ to arrive at a stronger straight exploded n-lacing. Since this would contradict our assumption, we conclude that

$$P_{(1+f)h}(h) \geq P_{(1+f)h}(2h).$$

Using property P4 on page 91, we conclude that

$$P_{(e+f)h}(h) > P_{(e+f)h}(2h).$$

For this to be the case, the function $P_{(e+f)h}$ has to be of type 2 and, therefore, as a direct consequence of property P2,

$$P_{(e+f)h}((e-1)h) > P_{(e+f)h}(eh).$$

This means that we can strengthen V by replacing an e and an f by an $e-1$ and an $f+1$, which is the contradiction we have been looking for.   □

Next, we prove the technical results about the function P_k that we refer to in the previous two proofs.

Lemma 7.10. *Let*

$$f : \mathbf{R} \to \mathbf{R} : x \mapsto \frac{1}{\sqrt{1+x^2}}$$

and

$$g_j : [-j, j] \to \mathbf{R} : x \mapsto f(x-j) + f(x+j),$$

where $j > 0$. Then the function g_j has the following properties:

- *It is symmetric about the y-axis.*
- *For $0 < j \leq 1/\sqrt{2}$, it has an absolute maximum at 0, and it is strictly increasing in the interval $[-j, 0]$.*
- *For $j > 1/\sqrt{2}$, it has a local minimum at 0 and a local maximum at some $x_j \in$ $]-j, 0[$. Furthermore, the function is strictly increasing in the interval $[-j, x_j]$ and strictly decreasing in the interval $[x_j, 0]$.*

Proof. Since f is symmetric about the y-axis, we conclude that g_j has the same property. Hence g_j has a local extremum at 0. It is a routine exercise to check that this extremum is a maximum for $0 < j \leq 1/\sqrt{2}$ and a minimum otherwise. Furthermore, the derivative of g_j at $-j$ is positive for all positive j. This implies that in the case $j > 1/\sqrt{2}$ the function g_j has at least one local maximum in the interval $]-j, 0[$. Therefore, it suffices to show that: (1) for $0 < j \leq 1/\sqrt{2}$, the derivative of g_j has no zero in the interval $]-j, 0[$; and (2) for $j > 1/\sqrt{2}$, the derivative of g_j has no more than one zero in the interval $]-j, 0[$.

The derivative of g_j is

$$[-j, j] \to \mathbf{R} : x \mapsto \frac{j-x}{(1+(j-x)^2)^{3/2}} - \frac{j+x}{(1+(j+x)^2)^{3/2}}.$$

If this function takes on the value zero at $x \in [-j, j]$, then x is also a solution of
the polynomial equation

$$(j + x)^2 (1 + (j - x)^2)^3 - (j - x)^2 (1 + (j + x)^2)^3$$
$$= (4j - 12j^5 - 8j^7)x + (24j^3 + 8j^5)x^3 + (-12j + 8j^3)x^5 - 8jx^7$$
$$= 0.$$

Note that, conversely, not every solution of this equation necessarily corresponds to
a zero of g'_j. We divide both sides of the equation by $4xj$ and thereby get rid of the
zero at the origin, which we are not interested in, as well as the common factor $4j$.
Following this, we substitute x^2 by the variable y. Then the number of solutions
to the equation above in the interval $] - j, 0[$ equals the number of solutions of the
equation

$$(1 - 3j^4 - 2j^6) + (6j^2 + 2j^4)y + (-3 + 2j^2)y^2 - 2y^3 = 0$$

in the interval $]0, j^2[$. Let's consider the left side of this equation as a function g in
the variable y. Clearly, g is a polynomial of degree 3.
Let V_l be the number of variations in sign in the sequence

$$g(0), g'(0), g''(0), g'''(0),$$

and let V_r be the number of variations in sign in the sequence

$$g(j^2), g'(j^2), g''(j^2), g'''(j^2).$$

Then the Budan-Fourier Theorem (see, for example, [19], Theorem 35) guarantees
that if $g(0) \neq 0$ and $g(j^2) \neq 0$, then the number of zeros of the equation $g(y) = 0$
in the interval $[0, j^2]$ is $V_l - V_r - 2l$, where l is some nonnegative integer.
The left-hand diagram in Figure 7.6 shows $g(0), g'(0), g''(0), g'''(0)$, considered as
functions in the variable j up to a point from which on their signs stay unchanged. It

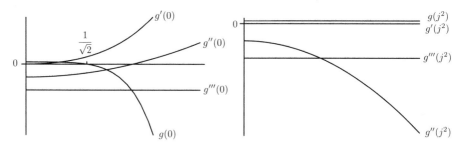

Figure 7.6. Plotting $g(0), g'(0), g''(0), g'''(0)$ as functions in j on the left, and
$g(j^2), g'(j^2), g''(j^2), g'''(j^2)$ as functions in j on the right.

is easy to check that $g(0)(j) = 0$ if and only if $j = 1/\sqrt{2}$. Also, note that $g'(0)$ is positive. For $0 < j < 1/\sqrt{2}$, we find that $V_l = 1$ and for $j > 1/\sqrt{2}$, we find that $V_l = 2$.
Similarly, the right-hand diagram in Figure 7.6 shows $g(j^2), g'(j^2), g''(j^2), g'''(j^2)$,
considered as functions in the variable j up to a point from which on their sign
stays unchanged. In particular, note that $g(j^2) = 1$ and $g'(j^2) = 0$. We find that

for all $j > 0$, $V_r = 1$. Hence, $V_l - V_r$ is 0 for $0 < j < 1/\sqrt{2}$ and 1 for $j > 1/\sqrt{2}$. This completes the proof of our result for $j < 1/\sqrt{2}$ and $j > 1/\sqrt{2}$. The case $j = 1/\sqrt{2}$ can easily be dealt with individually.   □

Lemma 7.11. *Let $k > 0$, and let P_k be the function*

$$[0, k] \to \mathbf{R} : x \mapsto \frac{1}{\sqrt{1 + x^2}} + \frac{1}{\sqrt{1 + (k - x)^2}},$$

in the interval $[0, k]$.

1. *If $k_0 \geq 2h > 0$ and $P_{k_0}(0) \geq P_{k_0}(h)$, then $P_k(0) > P_k(h)$ for all $k > k_0$.*
2. *If $k_0 \geq 4h > 0$ and $P_{k_0}(h) \geq P_{k_0}(2h)$, then $P_k(h) > P_k(2h)$ for all $k > k_0$.*

Proof. In the following, we may assume that P_k really has properties P1 and P2 listed on page 90. Please review these properties before you continue.

We start by proving part 1. Note that $P_{k_0}(0) \geq P_{k_0}(h)$ implies that P_{k_0} is of type 2. Therefore, by property P2, $k_0 > \sqrt{2}$.

Now, consider the function

$$Q : \mathbf{R} \to \mathbf{R} : k \mapsto P_k(0) - P_k(h) = 1 - \frac{1}{\sqrt{1 + h^2}} + \frac{1}{\sqrt{1 + k^2}} - \frac{1}{\sqrt{1 + (k - h)^2}}.$$

We want to show that this function is strictly increasing in the interval $]k_0, +\infty[$. Once we have proved this, we conclude that since, by assumption, $P_{k_0}(0) \geq P_{k_0}(h)$, we also have $P_k(0) > P_k(h)$ for all $k > k_0$, which proves the lemma.

The derivative Q' of Q is

$$\frac{k - h}{(1 + (k - h)^2)^{3/2}} - \frac{k}{(1 + k^2)^{3/2}}.$$

This means that $Q'(k) = 0$ if and only if the graph of the function

$$f : \mathbf{R} \to \mathbf{R} : k \mapsto \frac{k}{(1 + k^2)^{3/2}}$$

(see Figure 7.7) and its translate in the horizontal direction by a distance h intersect at the point $(k, Q'(k))$; see Figure 7.8.

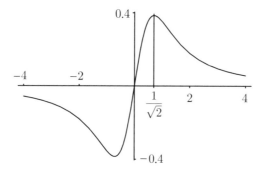

Figure 7.7. The graph of f.

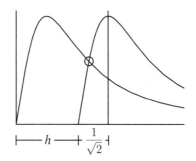

Figure 7.8. The unique positive zero of Q' corresponds to the point of intersection of the graph of f and its translate by a distance h.

It is easy to check that f has the following properties: (1) Its graph is symmetric about the origin. (2) It is strictly increasing in the interval $[0, 1/\sqrt{2}]$. (3) It is positive and strictly decreasing in the interval $[1/\sqrt{2}, +\infty[$. This implies that Q' has a unique positive zero $k_1 < h + 1/\sqrt{2}$ and is positive for $k > k_1$. Consequently, we can be sure that the function Q is strictly increasing for $k \geq h + 1/\sqrt{2}$; see Figure 7.8.

By assumption, $k_0 \geq 2h$. Hence, if $h \geq 1/\sqrt{2}$, then

$$k_0 \geq 2h = h + h \geq h + 1/\sqrt{2},$$

and we are done. On the other hand, we already showed that $k_0 > \sqrt{2}$. Hence, if $h < 1/\sqrt{2}$, then

$$k_0 > \sqrt{2} = 2/\sqrt{2} > h + 1/\sqrt{2}.$$

This completes the proof of part 1.

For part 2, let $k_0 \geq 4h > 0$ and $P_{k_0}(h) \geq P_{k_0}(2h)$. We want to show that $P_k(h) > P_k(2h)$ for all $k > k_0$. Let's consider the function

$$R : \mathbf{R} \to \mathbf{R} : k \mapsto P_k(h) - P_k(2h) =$$

$$\frac{1}{\sqrt{1 + h^2}} - \frac{1}{\sqrt{1 + 4h^2}} + \frac{1}{\sqrt{1 + (k - h)^2}} - \frac{1}{\sqrt{1 + (k - 2h)^2}}.$$

As in part 1, it suffices to show that this function is strictly increasing for $k > k_0$. Clearly, $R(k) = Q(k - h) + c$, for all $k \in \mathbf{R}$, where c is some constant. Therefore, since Q is strictly increasing for $k \geq h + 1/\sqrt{2}$, the function R is strictly increasing for $k \geq 2h + 1/\sqrt{2}$.

By assumption, $k_0 \geq 4h$. Hence, if $h \geq 1/(2\sqrt{2})$, then

$$k_0 \geq 4h = 2h + 2h \geq 2h + 1/\sqrt{2},$$

and we are done. On the other hand, we conclude as in part 1 that $k_0 > \sqrt{2}$. Hence, if $h < 1/(2\sqrt{2})$, then

$$k_0 > \sqrt{2} = 4/(2\sqrt{2}) > 2h + 1/\sqrt{2}.$$

This completes the proof of part 2.   □

Proof of Lemma 7.6. Let $n > 2$, and define the function

$$c_n : \mathbf{R} \to \mathbf{R} : h \mapsto Z(n, h) - C(n, h) = n - 2 + \frac{1}{\sqrt{1 + (n-1)^2 h^2}} - \frac{n-1}{\sqrt{1 + h^2}}.$$

Then, we need to show that the equation $c_n(h) = 0$ has exactly one solution h_n in the interval $]0, +\infty[$. Furthermore, we need to show that the function c_n is negative in the interval $]0, h_n[$ and positive in the interval $]h_n, +\infty[$.

First, note that $c_n(0) = 0$ and $\lim_{h \to +\infty} c_n(h) = n - 2 > 0$. Furthermore, it is easy to check that c_n has a local maximum at 0. This implies that c_n has at least one zero in the interval $]0, +\infty[$. It remains to show that there is no more than one zero in this interval. We can prove that this is the case by showing that c_n has no more than one local extremum in this interval.

The derivative of c_n is

$$(n-1)h \left(\frac{1}{(1 + h^2)^{3/2}} - \frac{(n-1)}{(1 + (n-1)^2 h^2)^{3/2}} \right).$$

If this derivative takes on the value zero at $h \in]0, +\infty[$, then h is also a solution of the polynomial equation

$$(1 + m^2 h^2)^3 - m^2 (1 + h^2)^3$$
$$= 1 - m^2 + (-3m^2 + 3m^4)h^4 + (-m^2 + m^6)h^6$$
$$= 0$$

where $m = n - 1$. Then, by Descartes' rule of signs (see, for example, [19], Corollary 35), the number of positive solutions of this equation is equal to the number of variations in sign of the nonzero coefficients of the polynomial on the left side of this equation minus some even, nonnegative integer. Since the coefficients are

$$-(m^2 - 1), 3m^2(m^2 - 1), m^2(m^4 - 1)$$

and $m > 1$, the number of variations in sign is 1, and the equation above has exactly one positive solution.   □

Proof of Lemma 7.9. Let $n > 3$, and define the function

$$c_n : \mathbf{R} \to \mathbf{R} : h \mapsto Z(n, h) - S(n, h) = \frac{n-3}{\sqrt{1 + h^2}} + \frac{1}{\sqrt{1 + (n-1)^2 h^2}} - \frac{n-2}{\sqrt{1 + 4h^2}}.$$

Then, we need to show that the equation $c_n(h) = 0$ has exactly one positive solution h_n. Furthermore, we need to show that the function c_n is negative in the interval $]0, h_n[$ and positive in the interval $]h_n, +\infty[$.

First, note that $c_n(0) = \lim_{h \to +\infty} c_n(h) = 0$. Furthermore, it is easy to check that c_n has a local maximum at $h = 0$. This means that c_n is negative for small h. Then

$$\lim_{h \to +\infty} h c_n(h) = n - 3 + \frac{1}{n-1} - \frac{n-2}{2} = \frac{n-4}{2} + \frac{1}{n-1} > 0.$$

This implies that the function c_n is positive for large h. We conclude that the equation $c_n(h) = 0$ has at least one positive solution. Furthermore, it is clear that if the equation has exactly one solution, then its graph will look roughly as the graph shown in the left diagram of Figure 7.9. If the equation has exactly two solutions, then the graph of c_n will look like one of the two graphs shown in the diagrams on the right. In particular, and this is very important, one of the solutions would correspond to a point on the x-axis which is also a local maximum or minimum of the graph.

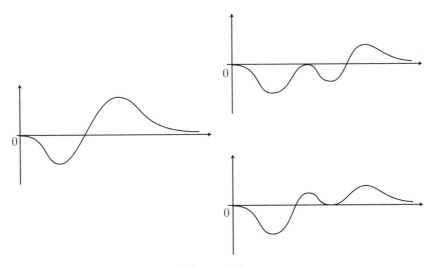

Figure 7.9.

We want to show that our equation $c_n(h) = 0$ has exactly one solution. Every solution of this equation is also a solution of

$$\left(\frac{n-3}{\sqrt{1+h^2}} + \frac{1}{\sqrt{1+(n-1)^2 h^2}}\right)^2 - \left(\frac{n-2}{\sqrt{1+4h^2}}\right)^2$$

$$= \frac{(n-3)^2}{1+h^2} + \frac{2(n-3)}{\sqrt{1+h^2}\sqrt{1+(n-1)^2 h^2}} + \frac{1}{1+(n-1)^2 h^2} - \frac{(n-2)^2}{1+4h^2}$$

$$= 0.$$

In turn, every solution of this equation is also a solution of

$$\left(\frac{2(n-3)}{\sqrt{1+h^2}\sqrt{1+(n-1)^2 h^2}}\right)^2 - \left(\frac{(n-2)^2}{1+4h^2} - \frac{1}{1+(n-1)^2 h^2} - \frac{(n-3)^2}{1+h^2}\right)^2$$

$$= 0.$$

Multiplying this equation by

$$(1+h^2)^2 (1+4h^2)^2 (1+(n-1)^2 h^2)^2$$

gives the following polynomial equation with the same solutions:

$$h^2(-144 + 240n - 148n^2 + 40n^3 - 4n^4)$$
$$+ h^4(-1008 + 2400n - 2380n^2 + 1260n^3 - 376n^4 + 60n^5 - 4n^6)$$
$$+ h^6(-1584 + 6672n - 10364n^2 + 8376n^3 - 3956n^4 + 1116n^5 - 176n^6 + 12n^7)$$
$$+ h^8(-720 + 4512n - 11048n^2 + 13960n^3 - 10145n^4 + 4404n^5 - 1126n^6$$
$$+156n^7 - 9n^8)$$
$$= 0.$$

As in the previous proof, we want to use Descartes' rule of signs to determine the number of positive solutions of this equation. If we consider the coefficients of the polynomial as functions in n, then it is easily checked that for $n > 5$ the signs of these polynomials are determined by the respective leading coefficient. This means that, in this case, the number of variations in sign of our polynomial in h is the number of variations in sign of the sequence $-4, -4, 12, -9$, which is 2. Clearly, every positive solution of the original equation is also a solution of the polynomial equation. However, the polynomial equation may have more positive solutions than the original equation. This means that the original equation that we are really interested in has either one or two positive solutions.

Let's assume that the original equation has two different positive solutions. Then we have already seen that one of the solutions corresponds to a point on the x-axis which is also a local maximum or minimum of the graph of c_n. However, this would also be true for the graph of the polynomial. Solutions like this are counted (at least) double by Descartes' rule of signs. This means that we would have at least three variations in sign, which is not the case. We conclude that our original equation has exactly one solution for $n > 5$. It is easy to check that this is also the case for $n = 4$ and $n = 5$.   □

7.5 Notes

Friction. In real shoes, friction will often result in the tension being not equal at all parts of a tied shoelace. This and a number of other considerations suggest that our interpretation of lacings as pulley systems is probably too simplistic a model to be of any real use in the analysis of the relative strengths of the different possible lacings of real shoes.

A related setup in which friction does not play too much of a role is the suturing of wounds. For a strength/tension analysis of three different suturing methods, see [22].

Shortest and strongest dense n-lacings. It is interesting to compare the problem of finding the shortest dense n-lacings with that of finding the strongest dense n-lacings. In the first problem the object is to minimize the sum

$$\sum_{i=1}^{2n} l_i,$$

where the l_is stand for the lengths of the different segments of a lacing. In the second problem we wish to maximize the sum

$$\sum_{i=1}^{2n} \frac{1}{l_i}.$$

Clearly, stated like this, we expect the solutions to both our problems to consist of short segments, because small l_i values give rise to lacings that are both short and strong. The actual solutions to these problems confirm this expectation to some extent, in that most segments of the crisscross and zigzag n-lacings are indeed very short. However, the zigzag n-lacings also contain one of the longest possible segments, which comes as a little bit of a surprise.

PEANUTS: © United Feature Syndicate, inc.

The Weakest Lacings

Out of curiosity, and for completeness' sake, I also used a computer to come up with some conjectures regarding the weakest lacings in the different classes of lacings under consideration. Figure 8.1 shows the picture that emerged from these experiments. As in the case of the longest shoelace problem, it was not my aim to prove or disprove all of these conjectures. However, some of our favorite lacings do show up in this diagram, and since it is fairly easy to prove a number of the characterizations of these lacings suggested by the computer experiments, we will do so in the following.

8.1 Summary of Results and Conjectures

The main result of this chapter is the following theorem:

Theorem 8.1 (Weakest). *The weakest simple, straight, simple-and-straight, dense-and-simple, and dense-and-simple-and-straight n-lacings for short and long shoes are as follows:*

(Simple) For short shoes and even n, the weakest simple n-lacings are the bowtie n-lacings. For short shoes and odd n, the weakest simple n-lacings are the zigzag n-lacings. For long shoes, the weakest simple n-lacings are the bowtie n-lacings.

(Straight) The weakest straight n-lacings are the superstraight n-lacings for even n and, for odd n, the n-lacings that contain n horizontals, one $(n - 1)$-diagonal, and $n - 1$ verticals.

(Simple-and-straight) The weakest simple-and-straight n-lacings are the simple-and-superstraight n-lacings for even n and, for odd n, the zigzag n-lacings.

(Dense-and-simple) The weakest dense-and-simple n-lacings are the zigzag n-lacings for short shoes and the crisscross n-lacings for long shoes.

(Dense-and-simple-and-straight) The weakest dense-and-simple-and-straight n-lacings are the zigzag n-lacings for short shoes and the star n-lacings for long shoes.

The individual results contained in the theorem and all our conjectures regarding the weakest lacings in the different classes of lacings under consideration are summarized in the Venn diagram in Figure 8.1. Here is a detailed account of how the individual diagrams in this Venn diagram should be interpreted. First, some of the diagrams are not diagrams of lacings, for example, the four diagrams at the top.

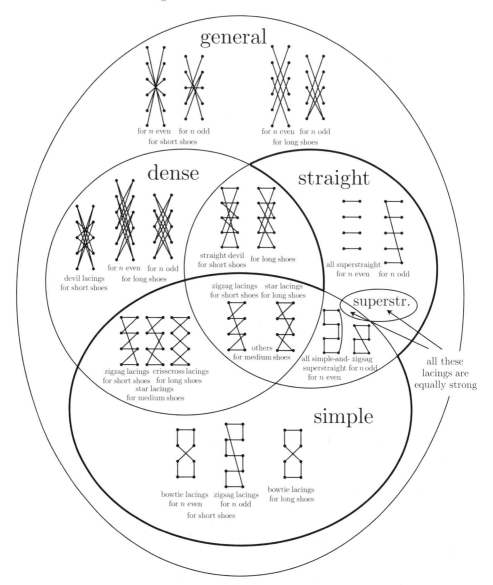

Figure 8.1. The weakest lacings in the different classes of n-lacings. The weakest lacings outside the thick border are only conjectures.

Any such diagram stands for all lacings that, apart from verticals, contain exactly the segments shown in the diagram (as well as horizontal and vertical mirror images of these lacings).

Second, to avoid a cluttered appearance, we show only the splitting up into the odd and even cases if we have not encountered a class of lacings previously. In particular, we display only a typical 'odd' representative for the devil lacings and the straight devil lacings.

Third, essentially, Theorem 8.1 corresponds to the diagrams enclosed in the thick loop. Note that Theorem 8.1 identifies only the weakest n-lacings in some of the classes of lacings for short and long shoes and does not say much about what happens for medium shoes. Here it is important to stress that it was exactly the absence of further extremal n-lacings for medium shoes, different from those for short and long shoes, that made the results of the previous chapters particularly surprising and appealing. Also, a closer look at our proofs in the previous chapters reveals that proving the absence of such extra extremal n-lacings really was what made some of these proofs tricky. If we had just been interested in the extremal lacings for long and short shoes, our proofs would have been much simpler. We only conjecture that the star n-lacings are the weakest dense-and-simple n-lacings for medium shoes and that there are many different weakest dense-and-simple-and-straight n-lacings for different medium shoes. In both these classes, it is easy to prove that there is at least one extra weakest n-lacing for some medium shoe.

Finally, we conjecture that in any of the remaining classes (general, dense, simple, dense-and-straight) there are only the two weakest solutions—one for short shoes and one for long shoes.

Proof of Theorem 8.1. Two simple observations about the contribution of horizontals and verticals towards the pulley sum of a lacing turn out to be very important in the following.

Observation 1 (Maximal Number of Verticals for Short Shoes): The contribution of a vertical to the pulley sum of a lacing stays unchanged and is zero as one shortens a shoe, whereas the contribution of a diagonal tends to 1 as the length of the shoe goes to zero. This means that if a lacing l has more verticals than a lacing k, then l will be weaker than k for short shoes. Consequently, the weakest n-lacings for short shoes in any of the classes under consideration contain the maximal possible number of verticals among the n-lacings in this class.

Observation 2 (Minimal Number of Horizontals for Long Shoes): The contribution of a horizontal to the pulley sum of a lacing stays unchanged and equals 1 as one lengthens a shoe, whereas the contribution of a diagonal that is not a horizontal tends to zero as the length of the shoe goes to infinity. This means that if a lacing l has less horizontals than a lacing k, then l will be weaker than k for long shoes. Consequently, the weakest n-lacings for long shoes in any of the classes under consideration contain the minimum possible number of horizontals among the n-lacings in this class.

We leave the proof of this theorem in the case $n = 2$ as an easy exercise and will assume $n > 2$ in the following. To keep this proof short, we will assume familiarity with the tools and arguments that we used in the proofs of the previous chapter. We prove the individual parts of this theorem out of order to be able to use arguments that build on each other.

Dense-and-simple. By Lemma 1.3, every simple n-lacing has at least two horizontals. The sum of the vertical lengths of the segments in a simple n-lacing is $2(n-1)$. This means that if a simple n-lacing contains only two horizontals, then all its other segments have vertical length 1. As a consequence of observation 2 and Lemma 7.3, this immediately implies that the crisscross n-lacing is the weakest dense-and-simple n-lacing for long shoes.

For short shoes, we can argue as in the proof of Theorem 7.1 by considering *dense-and-simple exploded n-lacings*, that is, sets of $2n$ nonnegative integers having the following properties: (1) Any such set does not contain more than n 0s. (2) The numbers in such a set add up to $2(n-1)$. (3) All elements in a set like this are less than or equal to $n-1$. We then choose h so small that the functions $P_{(v+v')h}$ are of type 1 (see page 90) for all possible choices of $0 \leq v, v' \leq n-1$ (which is the case if the function $P_{2(n-1)h}$ is of type 1). Now, if a dense-and-simple exploded n-lacing contains less than n 0s, then we can weaken it to a dense-and-simple exploded n-lacing by replacing two of its elements v, v' with $1 \leq v \leq v' < n-1$ by $v-1$ and $v'+1$. We conclude that a weakest dense-and-simple exploded n-lacing contains exactly n 0s. Similarly, we conclude that it contains $n-1$ 1s and that, therefore, it is the exploded zigzag n-lacing. As usual, we can now conclude that the zigzag n-lacings are the weakest dense-and-simple n-lacings for short shoes.

Dense-and-simple-and-straight. Since the zigzag n-lacings are straight and the weakest dense-and-simple n-lacings for short shoes, they are also the weakest dense-and-simple-and-straight n-lacings for short shoes.

To derive the weakest dense-and-simple-and-straight n-lacings for long shoes, we introduce *dense-and-simple-and-straight exploded n-lacings.* These are sets of n positive integers that add up to $2(n-1)$. Given a dense-and-simple-and-straight n-lacing, the corresponding dense-and-simple-and-straight exploded n-lacing is the set of vertical lengths of the nonhorizontal segments in the lacing. Clearly, every dense-and-simple-and-straight exploded n-lacing contains more than one 1. If such a set contains an element v greater than 2, then we can show for long enough shoes, in the by now usual way, that if we replace this v and one of the 1s by a $v-1$ and a 2, then we arrive at a weaker dense-and-simple-and-straight exploded n-lacing. This implies that the weakest dense-and-simple-and-straight exploded n-lacing consists only of 1s and 2s. In fact, it is clear that we need exactly two 1s and the rest 2s to get a sum of $2(n-1)$. However, the only dense-and-simple-and-straight n-lacings that give rise to this particular dense-and-simple-and-straight exploded n-lacing are the star n-lacings (see Lemma 4.5.1). As usual, we now conclude that the star n-lacings are the weakest dense-and-simple-and-straight n-lacings for long shoes.

Simple. We can argue as in the dense-and-simple case for long shoes to conclude that a weakest simple n-lacing for long shoes contains exactly two horizontals (the top and bottom horizontals) and that all its other segments have vertical length 1. Since the bowtie n-lacings are the only simple n-lacings containing exactly two horizontals and the minimal number of 1-diagonals (or, equivalently, the maximal number of 1-verticals), we conclude that the bowtie n-lacings are the weakest n-lacings for long shoes.

Since the bowtie n-lacings are simple and contain the maximum possible number of verticals, observation 2 guarantees that a weakest simple n-lacing for short shoes also contains this number of verticals. Let n be even and let l be an n-lacing that contains the maximum possible number of verticals. If all these verticals are 1-verticals, then these verticals are clearly the same verticals as those in the bowtie n-lacing. On the other hand, it is clear that if l contains a vertical v whose vertical length is greater than 1, then another vertical v' in the same column as v has to overlap with v in at least two eyelets. We may assume that the column in question is column A. Remember that since l is a simple n-lacing, eyelet A_1 and eyelet A_n cut l into two paths from A_1 to A_n that do not backtrack. Since v and v' overlap, this means that they are contained in different paths. This implies that v and v' overlap in exactly two consecutive eyelets A_i and A_{i+1}, as in the diagrams on the left in Figure 8.2. If they did overlap more, then any eyelet contained in the overlap that is not endpoint of either v or v' would not be contained in either path, because these paths do not backtrack.

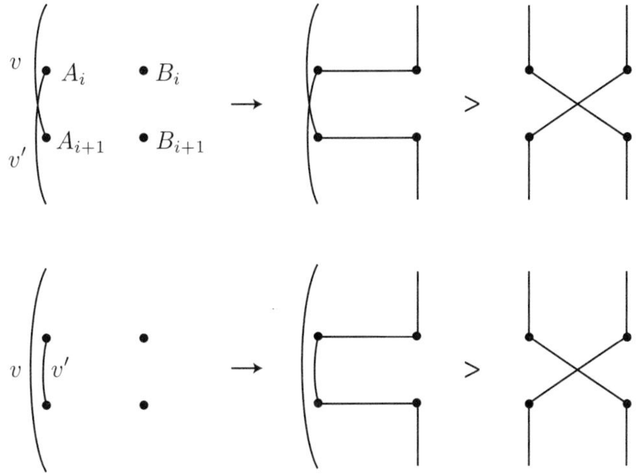

Figure 8.2.

Also because the two paths do not backtrack, for the eyelets B_i and B_{i+1} to be contained in these paths, the horizontals $A_i B_i$ and $A_{i+1} B_{i+1}$ have to be part of the lacing l and the paths have to continue as shown in the middle diagrams. But this means that we can weaken l by modifying l as indicated in the right diagram (the two horizontals get replaced by 1-diagonals and the verticals by other verticals).

We conclude that a weakest simple n-lacing for short shoes contains exactly the same verticals as the bowtie n-lacing. However, it is also easy to see that the only simple n-lacing that contains the same verticals as the bowtie n-lacing is the bowtie n-lacing. Summarizing, we conclude that the bowtie n-lacing is the weakest n-lacing for n even and short shoes.

Let n be odd. In this case the maximum possible number of verticals is $n-1$ and the number of diagonals in a lacing like this is $n+1$. Now, review the argument that we used to prove that zigzag n-lacings are the weakest dense-and-simple n-lacings for short shoes. It is a straightforward exercise to adapt this argument to prove that for short shoes the set consisting of n horizontals and one $n-1$ diagonal is the weakest among all sets of $n+1$ diagonals contained in an n-lacing. Finally, it is easy to see that the zigzag n-lacings are the only simple n-lacings that contain n horizontals, one $(n-1)$-diagonal and $n-1$ verticals. We conclude that the zigsag n-lacings are the weakest simple n-lacings for short shoes.

Straight and simple-and-straight. Since, by definition, a straight n-lacing contains all the horizontals, it is clear that, for even n, the superstraight n-lacings are the weakest straight n-lacings. Similarly, it is clear that the simple-and-superstraight n-lacings are the weakest simple-and-straight n-lacings. In the odd case, observe that the larger the vertical length of a diagonal, the less it contributes towards the pulley sum of a lacing. Since it is possible to build straight n-lacings that, apart from an $(n-1)$-diagonal and the horizontals, contain only verticals, every such

n-lacing is a weakest straight n-lacing. Since the zigsag n-lacings are exactly the n-lacings of this type that are also simple, the zigsag n-lacings are the weakest simple-and-straight n-lacings. □

8.2 Notes

As a consequence of Theorem 7.1 and Theorem 8.1, among the dense-and-simple n-lacings, the crisscross n-lacing is the strongest for short shoes and the weakest for long shoes, whereas the zigzag n-lacings are the strongest for long shoes and the weakest for short shoes. In addition, there is a unique value of the stretch of the underlying shoe for which the pulley sums of the zigzag and the crisscross n-lacings are equal and all other dense-and-simple n-lacings are weaker than both the crisscross and the zigzag n-lacings for this particular stretch of our mathematical shoe. This implies that there will be at least one dense-and-simple n-lacing different from these two types of n-lacings that is a weakest dense-and-simple n-lacing for this choice of stretch. Our numerical experiments suggest that apart from the crisscross and the zigzag n-lacings, the star n-lacings are the only other weakest dense-and-simple n-lacings.

A similar picture emerges for dense-and-simple-and-straight n-lacings. In this class the star n-lacings are the strongest lacings for short shoes and the weakest lacings for long shoes, whereas the zigzag n-lacings are the strongest ones for long shoes and the weakest ones for short shoes. In addition, there is a unique value of the stretch of the underlying shoe for which the pulley sums of the zigzag and the star n-lacings are equal and all other dense-and-simple-and-straight n-lacings are weaker than both the star and the zigzag n-lacings for this particular stretch of our mathematical shoe. As above, we conclude that there will be at least one dense-and-simple-and-straight n-lacing different from these two types of n-lacings that is a weakest dense-and-simple n-lacing. Our numerical experiments suggest that there will always be many different weakest dense-and-simple-and-straight n-lacings for medium shoes.

ROSE IS ROSE: © United Feature Syndicate, inc.

A

Related Mathematics

In this appendix, we first give a brief introduction to the so-called traveling salesman problems. These problems are close relatives of our shortest shoelace problems. In particular, we say a little bit about the important role that shortening rules, similar to the ones we used earlier on, play in the solution of these problems. We also describe the so-called shoelace formula for calculating the area of polygons.

A.1 Traveling Salesman Problems

If we are interested only in the shortest closed path in the plane that visits all $2n$ eyelets of a classical shoe exactly once, then the $2n$-eyelet counterpart of the lacing shown in Figure A.1 is what we are looking for. To see that this is the case, just note that this loop is definitely the shortest closed path in the plane that contains the four eyelets $A_1, B_1, A_n,$ and B_n.

Figure A.1. The shortest path that visits all eyelets.

This path is the solution of the so-called *traveling salesman problem* for our special configuration of points. In general, to solve the traveling salesman problem for a given set of points in the plane means to find a closed loop of minimal length that contains all points and to prove that this loop really has this minimality property. Just to give you an idea of what is possible here, Figure A.2 shows the solution to a traveling salesman problem involving more than 15,000 cities and villages in Germany. This solution is due to David Applegate, Robert Bixby, Vašek Chvátal, and William Cook; see [1].

Figure A.2. A solution to a traveling salesman problem that involves more than 15,000 cities and villages in Germany.

One of the most basic and interesting results about the solutions of a traveling salesman problem is that they never intersect themselves. The simplest way to prove this is to use one of our shortening rules. Figure A.3 shows how a self-intersecting loop containing a given set of points can be shortened using the shortening rule shown in Figure 5.7.

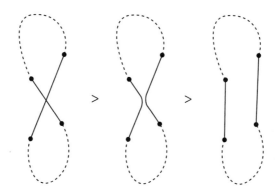

Figure A.3. A solution to a traveling salesman problem does not intersect itself.

As usual, we draw only that part of the loop solid to which we want to apply our shortening rule, and indicate by dotted curves how this part is tied into the overall path. Of course, it is possible that, by applying our shortening rule, we actually introduce new crossings somewhere else. However, it is also quite easy to show that if we just keep removing crossings like this, after a finite number of steps we will end up with a loop that does not intersect itself. This implies that a loop of minimal length containing all the points also has no self-intersections.

Our simple shortening rule and many more complicated shortening rules play important roles when it comes to solving complex traveling salesman problems.

Our lacing setup suggests the following generalization of the traveling salesman problem in its most basic form. Let C be a set of finitely many points in the plane that is the disjoint union of two sets A and B that contain the same number of points. Find the roundtrips of minimal length that visit every point in C once and do not visit three points in either of the two sets A or B consecutively; see also the results about lacing general n-shoes in Chapter 5.

See [17] for more information about solving traveling salesman problems.

A.2 The Shoelace Formula

The *shoelace formula* is a neat way of calculating the area of simply closed polygons in the plane. It can be summarized by way of a sufficiently general example as follows:

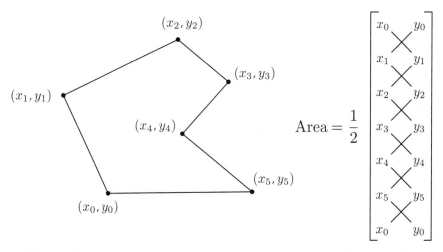

Figure A.4. Calculating the area of a polygon with the shoelace formula.

Of course, the crisscross pattern on the right is where the name "shoelace" formula comes from. It stands for the sum

$$|(x_0y_1+x_1y_2+x_2y_3+x_3y_4+x_4y_5+x_5y_0)-(x_1y_0+x_2y_1+x_3y_2+x_4y_3+x_5y_4+x_0y_5)|.$$

This means that the segments with negative slope represent the products in the first bracket and segments with positive slope the products in the second bracket.

In general, let $(x_0, y_0), (x_2, y_2), ..., (x_n, y_n) = (x_0, y_0)$ be the coordinates of the vertices of the polygon in the order that you come across these vertices as you traverse the polygon in the clockwise or counterclockwise direction, starting at the vertex with coordinates (x_0, y_0). Then,

$$\text{Area Polygon} = \frac{1}{2} \left| \sum_{i=0}^{n-1} x_i y_{i+1} - \sum_{i=0}^{n-1} x_{i+1} y_i \right| = \frac{1}{2} \left| \sum_{i=0}^{n-1} (x_i y_{i+1} - x_{i+1} y_i) \right|.$$

If we approximate a simply closed differentiable curve $(x(t), y(t))$, $t \in [0, 1]$, by polygons, then, in the limit, the sum on the right turns into the well-known formula for the area of such a curve

$$\frac{1}{2} \left| \int_0^1 (x(t) y'(t) - x'(t) y(t)) dt \right|.$$

Figure A.5. Vincent van Gogh (1853-1890), *A pair of shoes*, Paris, 1886, oilpaint on canvas 72 x 55 cm. Amsterdam, Van Gogh Museum (Vincent van Gogh Foundation).

B

Loose Ends

Collected in this second appendix are all kinds of curious and interesting facts about lacings of real shoes. Further bits and pieces about real lacings can be found in some of the footnotes and notes at the ends of chapters.

I have found that the mathematics of lacing shoes makes a great topic for an accessible and entertaining talk aimed at a very general audience. If you are interested in giving such a talk, the material in this appendix will be particularly useful.

B.1 History

A closer look at antique statues reveals that some of the ancient Romans and Greeks laced their sandals crisscross.

Figure B.1. A Roman sandal laced crisscross.

There is historical evidence of lacings of medieval shoes that did not form closed loops, as all the lacings used today do. Instead, the open zigzag lacing that we encountered in Section 5.4 was used, in which the two ends would be fastened individually with knots. Multiple shoelaces, such as those considered in Section 5.5, were also used. A first variation here included one shoelace per row of eyelets, another variation was one shoelace each for tying the top and bottom parts of a boot; see the middle and right diagrams in Figure B.2.

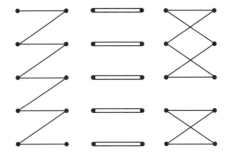

Figure B.2. Three other ways of lacing shoes used in the past.

B.2 Shoelace Superstitions

According to [18],

> shoelaces were often viewed with suspicion, partly because as a shoe style they were a comparatively late addition.... When the shoelace came undone accidentally, without being caught on anything and pulled out, this was interpreted as a true love thinking about you at that very moment. If the shoelaces became undone whilst walking, then your father loved you more than your mother. When the right shoelace came undone, then something good was being said about you and the opposite was true when the left shoelace was undone. A broken shoelace was thought to be bad luck.

PEANUTS: © United Feature Syndicate, inc.

B.3 Questions of Style

In the 18th century, the introduction of metal eyelets that prevented shoelaces from ripping through the upper of a shoe greatly enhanced the effectiveness of lacing shoes (and corsets), as now much greater force than before could be used to pull the two sides of a shoe together.

Subsequently, two classical types of gentlemen's shoes emerged—the *Oxford* and the *Derby*. One of the main distinguishing feature of these two kinds of shoes is the way the opening at the top is created. In Oxfords this is basically just a straight incision. In Derbys the opening consists of two flaps that can be opened wide; see Figure B.3.

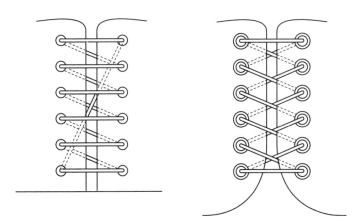

Figure B.3. Traditionally, Oxfords, on the left, and Derbys, on the right, were laced zigzag and crisscross, respectively.

The following quote is taken from the excellent book *Handmade Shoes for Men* ([29], p. 181). It tells you all you need to know when it comes to properly lacing these types of shoes.

> Fabric laces are best for made-to-measure gentlemen's shoes: round laces will suit elegant shoes, and either round or ribbon laces look good with sturdy shoes. Both kinds of lace are made by machine. Round laces are cylindrical, with a core made of cotton waste or hemp yarn. Ribbon laces usually have no core. To facilitate the threading of the laces through the eyelets, the ends have cylindrical metal tags fitted.
>
> The color of the shoelace must always suit the color of the upper; at most it may be a shade darker, but certainly no lighter. If the upper of the shoe is a combination of colors, then the laces should match the darker color. Their length depends on the number of eyelets; for shoes with five to six eyelets, the ideal length of the laces is about 30 inches (75 cm); for shoes with up to four eyelets it is 2 feet (60 cm). In this way a bow of the right size can be tied when the laces have been threaded through. Boots need laces 4 feet (120 cm) long.
>
> Both Oxfords and Derbys have their characteristic styles of lacing (Oxfords zigzag and Derbys crisscross), but it can be varied as you like. The methods are no longer bound to the original model. However tradition will always play a part, and most wearers will prefer to lace their shoes in the usual way for a particular style. Many people will recognize this care for what is customary as the mark of a gentleman.

Also, remember that there are two zigzag n-lacings, one the vertical mirror image of the other. So, if you prefer to lace your shoes zigzag, then you may want to consider lacing your right shoe with one and your left shoe with the second zigzag n-lacing to create a symmetrical appearance.

B.4 Fashion

You will find new models of shoes on the shelves of shoe shops every year. However, the basic types of lacings used to lace these shoes for the most part remain the standard ones, and the decorative potential of other lacings remains largely undeveloped. I find this very surprising in an area in which usefulness or health considerations are often considered not to be nearly as important as looks. Of course, there are exceptions. For example, a couple of years ago it became fashionable to use two or three shoelaces of different colors to lace sneakers.

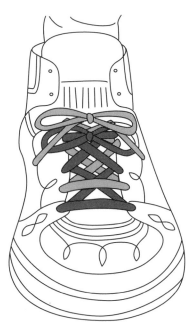

Figure B.4. *LA Gear* used to sell sneakers laced with two or more shoelaces of different colors.

Superstraight lacings are quite often used for display purposes in shoe shops. Here the shoes get laced in such a way that all verticals are hidden below the upper of the shoes and only the horizontals are visible. For this reason, superstraight lacings are also referred to as *blind* lacings; see the top left diagram in Figure 3.7 on page 28. If there is a gap between the two flaps on top of the shoe, this sometimes creates the illusion that the horizontals are not connected in any way.

In general, modifying the way a lacing is arranged over and under the upper of a shoe presents us with even more scope for variations. For example, Figure B.5 shows four decorative ways of lacing the crisscross lacing. According to the information on his website [10], Monte Fischer even managed to patent the method shown on the top left. Also have a look at the decorative lacings on page 58 and those listed on Ian Fieggen's website [7].

Figure B.5. A number of different decorative ways to lace crisscross.

B.5 What Is the Best Way to Lace Your Shoes?

As we have seen, the answer to this question depends on what we want "best" to mean. In this book, we have given answers to this question for those two interpretations of best that are most easily quantifiable: *best equals shortest* and *best equals strongest*. However, in practice, we may be even more interested in interpretations such as *best equals easiest to learn* or *best equals easiest to execute*. Actually, the fact that the classical lacings are definitely just about the easiest imaginable lacings to learn and to execute may be the main reason for their overwhelming popularity.

So, what is the best way to lace your shoes? Personally, I think that for most people the crisscross lacings are the best lacings. Apart from being very strong, short, and extremely easy to learn and execute, they also have an edge over most other lacings: when you are finished with lacing a crisscross lacing, you automatically end up with two ends of equal length, something very desirable that does not happen with most other lacings.

B.6 Foot Problems and Lacings

While the crisscross lacings may well be the ideal lacings for most people, depending on the shape of your foot, a different lacing method may actually help to make your shoe fit much better and may even help alleviate foot problems. A very informative interview with the orthopedic surgeon Carol Frey about this podiatric side of lacings can be found in [4]. Examples of lacings recommended in this article are shown in Figure B.6.

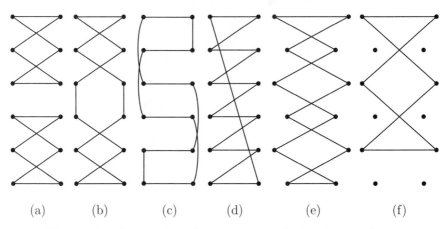

(a)   (b)   (c)   (d)   (e)   (f)

Figure B.6. Some lacings that are used to alleviate foot problems.

If you have a narrow heel and wide forefoot, Frey recommends lacing (a) on the left. Here you are supposed to tie the top shoelace more tightly than the bottom one. Similarly, if you have a high arch or pain from a tendon injury, Frey recommends to try leaving a space in the crisscross lacing as in (b) to alleviate the pressure. Alternatively, you may want to try a superstraight lacing like the serpent lacing (c). Unlike the crisscross lacings, these kinds of lacings create no pressure points where laces overlap. If you have problems with your toes, it is possible to adjust the height of the front part of your shoe using a zigzag lacing: provided there is enough friction to prevent the shoelace from sliding freely around the eyelets, just pulling on the free end of the lace that forms the long diagonal segment of the lacing will increase this height. To achieve a very snug fit, many modern sneakers feature an array of eyelets in which the horizontal spacing may vary from row to row as, for example, in the array of eyelets in diagram (e). According to Frey, if your feet are very narrow, you can improve the fit of such shoes by using only the pairs of eyelets that are furthest apart (f). This pulls the sides of the shoe tightly across the top part of your foot. Similarly, if your feet are very wide, use only those pairs of eyelets that are closest together.

The armed forces of some countries have a preference for straight lacings such as the zigzag and the serpent lacings because these straight lacings are both very easy to cut open with a knife in case a boot has to be removed from a wounded soldier's foot. For example, it was mentioned in [28] that the Canadian armed forces prefer serpent lacings.

In [23] Charles Schulz tells a shoelace story that actually happened in World War II. Soldiers were developing foot problems because they would not take off their shoes several days in a row. To prevent this from happening, the soldiers were ordered to use one pair of boots laced crisscross and another pair of boots laced zigzag on alternate days. The *Peanuts* cartoon reproduced below was inspired by this story.

B.7 What Is the Best Way to Tie Your Shoelaces?

You may think that the book you are holding in your hands is the first book dedicated to shoelaces. This is not so. In fact, if you check in the kids department of any bookshop, you may find several different shoelace books. These books try to teach kids to lace their shoes and tie their shoelaces, with the emphasis usually on the tying of shoelaces the way most of us do. Examples of such books are [3], [6], [30], and [31].

Writing books that teach kids to tie their shoelaces may seem a bit pointless; after all, everybody is an expert in the art of shoelace tying, and every parent should be able to pass on this important skill without any help. It may come as a surprise that this is not the case; although we all seem to be tying our shoelaces the same way, a very large number of people, possibly even the majority, do tie their laces much worse than the rest. Let me explain what I mean by this.

Most of us tie our shoelaces by first tying a straightforward half-knot, followed by a second more elaborate half-knot. The end result looks roughly as the knot in Figure B.7. As you can see, the second half-knot is basically the same as the first.

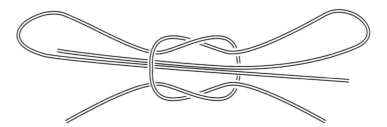

Figure B.7. A typical knot used for tying shoelaces.

This becomes even clearer once you remove the two loops by poking two fingers into the loops and pulling until the loose ends slide free, as shown in Figure B.8.

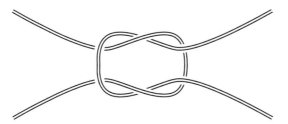

Figure B.8. A reef knot.

However, there are two different ways of putting one half-knot on top of another, and, after removing the loops, you may also end up with the knot shown in Figure B.9.

These two knots are called the *reef knot* and the *granny knot*. The granny knot results from tying two identical half-knots on top of each other. On the other hand, the two half-knots that form a reef knot are mirror images of each other.

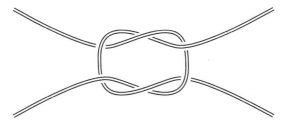

Figure B.9. A granny knot.

It turns out and every sailor knows this that reef knots are very stable and granny knots are not. This has the effect that if you happen to be a granny-knot person, your shoelaces are much more likely to come undone than if you are a reef-knot person. If you discover that you are a granny-knot person, don't despair! To turn yourself into a reef-knot person, all you have to do when you tie your shoelaces is change the orientation of the first half-knot and leave the complicated part of the tying action unchanged. Very simple!

I should mention that all the shoelace books that I have seen actually teach tying a reef knot. However, in most of them the diagrams that are supposed to illustrate the action of tying shoelaces confuse more than they help. An exception is [3].

There are a couple of other popular ways of tying shoelaces that make it harder for them to come undone. One consists of just tying a third half-knot into the loops, another is a variation of the reef knot taught in medical schools for suturing purposes; see Figure B.10.

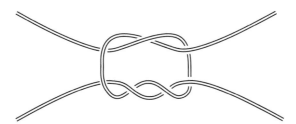

Figure B.10. A knot used for suturing.

As you can see, the first half-knot, which has one twist, gets replaced by a half-knot that has two twists. Even by itself such a doubly twisted half-knot does not easily come undone. Therefore, this suturing way of tying shoelaces is also a great way to teach kids to tie their laces. After all, one of the main problems for kids is that the first half-knot will be all loose before they manage to fasten the second half-knot.

For a lot more information about real shoelace knots and lacings, see Ian Fieggen's webpage [7]. Even the famous *Ashley Book of Knots* [2] features a small but interesting section on lacing and tying shoes. A quick Internet search yields further weird and wonderful facts about shoelaces, such as a patented pneumatic shoelacing machine. Finally, if you enjoyed learning about the mathematics of lacing, then you will almost certainly also be interested in the beautiful mathematics of tying a tie; see [8] and [9].

DILBERT: © Scott Adams/Dist. by United Feature Syndicate, inc.

References

1. Applegate, D., Bixby, R., Chvátal, V., and Cook, W. *The Traveling Salesman Problems Page* at www.tsp.gatech.edu/d15sol.
2. Ashley, C.W. *The Ashley Book of Knots*. Faber and Faber, London–Boston, 2000.
3. Casey, M. *Red Lace, Yellow Lace*. Barron, New York, 1996.
4. Cox, R. Lacing lessons. *University of Southern California Chronicle*, March 27, 1995, p. 12.
5. Dehner Company. *Dehner Lace Instructions* at www.dehner.com/dehner/lace.html.
6. Faulkner, K. *Loose Lace*. The Five Mile Press, Noble Park, Victoria, Australia, 1996.
7. Fieggen, I. *Ian's Shoelace Site* at www.fieggen.com/shoelace.
8. Fink, T.M. and Mao, Y. Designing tie knots by random walks. *Nature* 398 (4 March 1999), 31–32.
9. Fink, T.M. and Mao, Y. *85 Ways to Tie a Tie*. Fourth Estate, London, 1999.
10. Fisher, M. *Patented Double Helix Lacing* at www.lukefisher.com/lacing.
11. Gale, D. *Tracking the Automatic Ant and Other Mathematical Explorations*. Springer-Verlag, New York, 1998.
12. Gardner, M. *Martin Gardner's Sixth Book of Mathematical Games from* Scientific American. W.H. Freeman and Co., San Francisco, 1971.
13. Halton, J.H. The shoelace problem. Department of Computer Science Technical Report No. 92–032, University of North Carolina at Chapel Hill, 1992.
14. Halton, J.H. The shoelace problem. *The Mathematical Intelligencer* 17 (1995), 37–41.
15. Hildebrandt, S. and Tromba, A. *The Parsimonious Universe*. Copernicus, New York, 1996.
16. Isaksen, D.C. Shortest shoelaces. *Mathematics Magazine* 73 (2000), 60–61.
17. Johnson, D.S. and McGeoch, L.A. The traveling salesman problem: a case study. *Local Search in Combinatorial Optimization*, 215–310, Wiley-Intersci. Ser. Discrete Math. Optim., Wiley, Chichester, 1997.
18. Kippen, C. *The History of Shoes*. Online book available at www.curtin.edu.au/curtin/dept/physio/podiatry/history.html.
19. MacDuffey, C.C. *Theory of Equations*. John Wiley & Sons, New York, 1954.
20. Misiurewicz, M. Lacing irregular shoes. *The Mathematical Intelligencer* 18 (1996), 32–34.
21. Polster, B. What is the best way to lace your shoes? *Nature* 420 (5 December 2002), p. 476.
22. Rubinstein, C. and Russel, W.J. Wound closure and suturing patterns: a vector analysis of suture tension. *Aust. N.Z. J. Surg.* 62 (1992), 733–737.
23. Schulz, C.M. *You Don't Look 35, Charlie Brown!* Holt, Rinehart and Winston, New York, 1985.

24. Sloane, N.J.A. *On-Line Encyclopedia of Integer Sequences.* Online resource available at www.research.att.com/~njas/

25. Smith, G.A. *Shoelace Knots—the Ins and Outs* at www.u.arizona.edu/~gasmith/knots/knots.html.

26. Steinhaus, H. *One Hundred Problems in Elementary Mathematics.* Pergamon Press, Oxford, 1963.

27. Stewart, I. Arithmetic and old lace. *Scientific American,* July 1996, 78–80.

28. Stewart, I. Feedback (to [27]). *Scientific American,* December 1996, p. 86.

29. Vass, L. and Molnár, M. *Handmade Shoes for Men.* Könemann, Cologne, Germany, 1999.

30. *Walt Disney's Goofy Shoelace Book.* Tuffy Books, New York, 1988.

31. *Whose Shoes Are These?* Parragon, Bath, United Kingdom, 2001.

Index

Titles in This Series

For a complete list of titles in this series, visit the
AMS Bookstore at **www.ams.org/bookstore/**.